RECENT ADVANCES IN FINANCIAL ENGINEERING

Proceedings of the
2008 Daiwa International Workshop on
Financial Engineering

RECENT ADVANCES IN FINANCIAL ENGINEERING

Proceedings of the
2008 Daiwa International Workshop on
Financial Engineering

Otemachi Sankei Plaza, Tokyo, Japan
4 – 5 August 2008

editors

Masaaki Kijima
Tokyo Metropolitan University, Japan

Masahiko Egami
Kyoto University, Japan

Kei-ichi Tanaka
Tokyo Metropolitan University, Japan

Yukio Muromachi
Tokyo Metropolitan University, Japan

World Scientific

NEW JERSEY • LONDON • SINGAPORE • BEIJING • SHANGHAI • HONG KONG • TAIPEI • CHENNAI

Published by

World Scientific Publishing Co. Pte. Ltd.

5 Toh Tuck Link, Singapore 596224

USA office: 27 Warren Street, Suite 401-402, Hackensack, NJ 07601

UK office: 57 Shelton Street, Covent Garden, London WC2H 9HE

British Library Cataloguing-in-Publication Data
A catalogue record for this book is available from the British Library.

RECENT ADVANCES IN FINANCIAL ENGINEERING
Proceedings of the 2008 Daiwa International Workshop on Financial Engineering

ISBN-13 978-981-4273-46-6
ISBN-10 981-4273-46-5

Printed in Singapore by World Scientific Printers

We would like to express our special thanks to him in a memorable way. ɔook is devoted to Yuri Kabanov's 60th anniversary.

February, 2009

aki Kijima, Tokyo Metropolitan University/ Kyoto University
hiko Egami, Kyoto University
hi Tanaka, Tokyo Metropolitan University
Muromachi, Tokyo Metropolitan University

PREFACE

This book is the Proceedings of *Daiwa International Wo* *Engineering* held in Summer, 2008, which has been held in To 2004, in order to exchange new ideas in financial engineering

First of all, we would like to emphasize that the worksh Daiwa Securities Group, who donated the chair, named "Fin System," to Graduate School of Economics, Kyoto Univers then, collaborations between Daiwa Securities Group and us ious ways. Also, we would like to mention that the worksl nized by Graduate School of Social Sciences and Tokyo Metr (TMU). Recently, TMU has been granted by the Japan Societ of Sciences Program, which also supports this workshop.

In each year, various kinds of interesting studies are pres shop by many researchers from various countries, not only also from industry. This workshop serves as a bridge bet searchers and practitioners. However, the speakers did not n available before. So, we have been contemplating an approp order to make better use of the presentations; fortunately, th published by World Scientific Publishing Co. 12 papers are inc that almost cover the topics presented at the 2008 Daiwa I shop. These papers are addressing state-of-the-art techniques neering.

We would like to express our plenty of gratitude to tho their papers to this proceedings and those who helped us ki these papers. We also thank Professor Jiro Akahori, Ritsumeil his kind advice about publishing the proceedings, Mr. Satoshi manuscripts and Ms. Zhang Ruihe and Ms. Kakarlapudi Shal Scientific Publishing Co. for their kind assistance in publishing

Last but not least, we celebrated Professor Yuri Kabanov' the last year. Professor Kabanov has been contributing not only mathematical finance but also to the great success of the Daiwa chair. He was a visiting professor of the chair twice, from April

2008 Daiwa International Workshop on Financial Engineering

Date

August 4–5, 2008

Place

Otemachi Sankei Plaza, Tokyo, Japan

Organizer

Daiwa Securities Group Chair of Financial Engineering Graduate School of Economics, Kyoto University
Tokyo Metropolitan University
Japan Society for Promotion of Science's Program for Grants-in-Aid for Scientific Research (B) #18310104

Sponsor

Daiwa Securities Group / 大和証券グループ

Program Committee

Masaaki Kijima, Kyoto University, Chair
Masahiko Egami, Kyoto University
Katsumasa Nishide, Yokohama National University
Yusuke Osaki, Kyoto University
Takashi Shibata, Tokyo Metropolitan University
Keiichi Tanaka, Tokyo Metropolitan University

Program

August, 4 (Monday)

9:45–10:00 Toshiro Mutoh, Chaiman of the Institute, Daiwa Institute of Research
Opening Address

Chair: Keiichi Tanaka

10:00–10:45 Damien Lamberton, Universite Paris-Est Marne-la-Vallee
American Option Prices in Exponential Levy Models

10:45–11:00 Morning Coffee

11:00–11:30 Michi Nishihara, Osaka University
Investment Game with Debt Financing (with Takashi Shibata)

11:30–12:00 Katsumasa Nishide, Yokohama National University
Compensation Measures for Alliance Formation: A Real Options Analysis (with Yuan Tian)

12:00–13:15 Lunch

Chair: Naoki Makimoto

13:15–14:00 Masaaki Kijima, Kyoto University, Tokyo Metropolitan University
Estimation of the Local Volatility of Discount Bonds Using the Market Prices of Coupon Bond Options (with Hajime Fujiwara and Katsumasa Nishide)

14:00–14:30 Katsushige Sawaki, Nanzan University
The Valuation of Callable Financial Commodities with Two Stopping Boundaries

14:30–15:00 Yoshio Miyahara, Nagoya City University
Option Pricing Based on the Geometric Stable Processes and the Minimal Entropy Martingale Measures (with Naruhiko Moriwaki)

15:00–15:15 Afternoon Coffee

Chair: Masahiko Egami

15:15–16:00 Jin-Chuan Duan, National University of Singapore
Co-Integration in Crude Oil Components and the Pricing of Crack Spread Options (with Annie Theriault)

16:00–16:30 Masato Ubukata, Osaka University
A Test for Dependence and Covariance Estimator of Market Microstructure Noise (with Kosuke Oya)

August, 5 (Tuesday)

Chair: Katsushige Sawaki

9:45–10:30 Jean-Pierre Fouque, University of California Santa Barbara
Multiname and Multiscale Default Modeling (with R. Sircar and K. Solna)

10:30–10:45 Morning Coffee

10:45–11:15 Hoi Ying Wong, Chinese University of Hong Kong
Quanto Options under Double Exponential Jump Diffusion

11:15–11:45 Kenji Wada, Keio University
Uninsurable Risk, Bond Pricing and Real Interest Rates: An Investigation of UK Indexed Bonds (with David Barr and Parantap Basu)

11:45–13:15 Lunch

Chair: Yuji Yamada, Naoki Makimoto

13:15–14:00 Yuri Kabanov, Université de Franche-Comté
Hedging of American and European Options under Transaction Costs

14:00–14:30 Emmanuel Denis, University of Besançon
The Leland Approximations for European Options

14:30–15:00 Yoshihiko Sugihara, Bank of Japan
The New Variation Method for Volatility Risk Premium

15:00–15:15 Afternoon Coffee

Chair: Yusuke Osaki

15:15–15:45 Keiichi Tanaka, Tokyo Metropolitan University
Time Preference Induced by Risk Aversion (with Shinsuke Ikeda)

15:45–16:15 Jie Qin, Ritsumeikan University
Regret Aversion and Information Cascade in a Sequential Trading Model

16:15–16:30 Shuzo Nishimura, Vice-President, Kyoto University
Closing Address

Professor Yuri Kabanov

in 2008 Daiwa International Workshop

CONTENTS

Mean Square Error for the Leland–Lott Hedging Strategy

Moussa Gamys and Yuri Kabanov

Laboratoire de Mathématiques, Université de Franche-Comté
E-mail: ykabanov@univ-fcomte.fr

The Leland strategy of approximate hedging of the call-option under proportional transaction costs prescribes to use, at equidistant instants of portfolio revisions, the classical Black–Scholes formula but with a suitably enlarged volatility. An appropriate mathematical framework is a scheme of series, i.e. a sequence of models M_n with the transaction costs coefficients k_n depending on n, the number of the revision intervals. The enlarged volatility $\widehat{\sigma}_n$, in general, also depends on n. Lott investigated in detail the particular case where the transaction costs coefficients decrease as $n^{-1/2}$ and where the Leland formula yields $\widehat{\sigma}_n$ not depending on n. He proved that the terminal value of the portfolio converges in probability to the pay-off. In the present note we show that it converges also in L^2 and find the first order term of asymptotics for the mean square error. The considered setting covers the case of non-uniform revision intervals. We establish the asymptotic expansion when the revision dates are $t_i^n = g(i/n)$ where the strictly increasing scale function $g : [0,1] \to [0,1]$ and its inverse f are continuous with their first and second derivatives on the whole interval or $g(t) = 1 - (1-t)^\beta, \beta \geq 1$.
Key words: Black–Scholes formula, European option, transaction costs, Leland–Lott strategy, approximate hedging

1. Introduction

1.1 Formulation of the Main Result

To fix the notation we consider the classical Black–Scholes model, already under the martingale measure and with the maturity $T = 1$. So, let $S = (S_t)$, $t \in [0,1]$, be a geometric Brownian motion given by the formula

$$S_t = S_0 e^{\sigma W_t - \frac{1}{2}\sigma^2 t}$$

and satisfying the linear equation

$$dS_t = \sigma S_t dW_t$$

with a standard Wiener process W and constants S_0, $\sigma > 0$. Let $C(t, x)$ be the solution, in the domain $[0, 1]\times]0, \infty[$, of the Cauchy problem

(1) $$C_t(t, x) + \frac{1}{2}\sigma^2 x^2 C_{xx}(t, x) = 0, \qquad C(1, x) = (x - K)^+,$$

where $K > 0$. The function $C(t, x)$ admits an explicit expression and this is the famous Black–Scholes formula:

(2) $$C(t, x) = C(t, x, \sigma) = x\Phi(d) - K\Phi(d - \sigma\sqrt{1-t}), \qquad t < 1,$$

where Φ is the Gaussian distribution function with the density φ,

(3) $$d = d(t, x) = d(t, x, \sigma) = \frac{1}{\sigma\sqrt{1-t}} \ln\frac{x}{K} + \frac{1}{2}\sigma\sqrt{1-t}.$$

Define the process

(4) $$V_t = C(0, S_0) + \int_0^t C_x(u, S_u) dS_u.$$

In the Ito formula for $C(t, S_t)$ the integral over dt vanishes. Hence, $V_t = C(t, S_t)$ for all $t \in [0, 1]$. In particular, $V_1 = (S_1 - K)^+$: at maturity the value process V replicates the terminal pay-off of the call-option.

Modelling assumptions of the above formulation are, between others: frictionless market and continuous trading. The latter is a purely theoretical invention. Practically, an investor revises the portfolio at certain dates t_i and keeps $C_x(t_i, S_{t_i})$ units of the stock until the next revision date t_{i+1}. The model becomes more realistic if the transactions are charged proportionally to their volume. The portfolio strategy suggested by Leland [6] generates the value process

(5) $$V_t^n = \widehat{C}(0, S_0) + \int_0^t \sum_{i=1}^n H_{t_{i-1}}^n I_{]t_{i-1}, t_i]}(u) dS_u - \sum_{t_i < t} k_n S_{t_i} |H_{t_i}^n - H_{t_{i-1}}^n|,$$

where $H_{t_i}^n = \widehat{C}_x(t_i, S_{t_i})$, $t_i = i/n$, the positive parameter $k_n = k_0 n^{-1/2}$ is the transaction costs coefficient, and $\widehat{C}(t, x)$ is the solution of (1) with σ replaced by $\widehat{\sigma} > 0$ with

(6) $$\widehat{\sigma}^2 = \sigma^2 + \sigma k_0 \sqrt{8/\pi}.$$

That is $\widehat{C}(t, x) = C(t, x, \widehat{\sigma})$ and for such a strategy there is no need in a new software: traders can use their old one, changing only one input parameter, the volatility.

In his paper Leland claimed, without providing arguments, that V_1^n converges to $V_1 = (S_1 - K)^+$ in probability as $n \to \infty$. This assertion was proven by Lott in his thesis [7] and we believe that the result could be referred to as the Leland–Lott theorem. In fact, V_1^n converges also in L^2 and the following statement gives the rate of convergence:

Theorem 1.1. *The mean square approximation error of the Leland–Lott strategy has the following asymptotics:*

$$E(V_1^n - V_1)^2 = A_1 n^{-1} + o(n^{-1}), \qquad n \to \infty, \tag{7}$$

where the coefficient

$$A_1 = \int_0^1 \left[\frac{\sigma^4}{2} + \sigma^3 k_0 \sqrt{\frac{2}{\pi}} + k_0^2 \sigma^2 \left(1 - \frac{2}{\pi}\right) \right] \Lambda_t dt \tag{8}$$

with $\Lambda_t = ES_t^4 \widehat{C}_{xx}^2(t, S_t)$. Explicitly,

$$\Lambda_t = \frac{K^2}{2\pi\widehat{\sigma}\sqrt{1-t}\sqrt{2\sigma^2 t + \widehat{\sigma}^2(1-t)}} \exp\left\{ -\frac{\left(\ln\frac{S_0}{K} - \frac{1}{2}\sigma^2 t - \frac{1}{2}\widehat{\sigma}^2(1-t)\right)^2}{2\sigma^2 t + \widehat{\sigma}^2(1-t)} \right\}. \tag{9}$$

The main result of this note is slightly more general and also covers a model with non-uniform grids given as follows.

Let f be a strictly increasing smooth function on $[0,1]$ with $f(0) = 0$, $f(1) = 1$ and let $g := f^{-1}$ denote its inverse. For each fixed n we define the revision dates $t_i = t_i^n = g(i/n)$, $1, ..., n$. The enlarged volatility now depends on t and is given by the formula

$$\widehat{\sigma}_t^2 = \sigma^2 + \sigma k_0 \sqrt{8/\pi}\sqrt{f'(t)} \tag{10}$$

while the function $\widehat{C}(t,x)$ given by the formula

$$\widehat{C}(t,x) = x\Phi(\rho_t^{-1}\ln(x/K) + \rho_t/2) - K\Phi(\rho_t^{-1}\ln(x/K) - \rho_t/2), \qquad t < 1,$$

with $\rho_t^2 = \int_t^1 \widehat{\sigma}_s^2 ds$ solves the Cauchy problem

$$\widehat{C}_t(t,x) + \frac{1}{2}\widehat{\sigma}_t^2 x^2 \widehat{C}_{xx}(t,x) = 0, \qquad \widehat{C}(1,x) = (x-K)^+, \tag{11}$$

The following bounds are obvious:

$$\sigma^2(1-t) \le \rho_t^2 \le \sigma^2(1-t) + \sigma k_0 \sqrt{8/\pi}(1-t)^{1/2}(1-f(t))^{1/2}.$$

Assumption 1: $g, f \in C^2([0,1])$.

Assumption 2: $g(t) = 1 - (1-t)^\beta, \beta \geq 1$.

Note that in the second case where $f(t) = 1 - (1-t)^{1/\beta}$ the derivative f' for $\beta > 1$ explodes at the maturity date and so does the enlarged volatility.

Theorem 1.2. *Under any of the above assumptions*

$$E(V_1^n - V_1)^2 = A_1(f)n^{-1} + o(n^{-1}), \qquad n \to \infty, \tag{12}$$

where the coefficient

$$A_1(f) = \int_0^1 \left[\frac{\sigma^4}{2}\frac{1}{f'(t)} + k_0\sigma^3\sqrt{\frac{2}{\pi}}\frac{1}{\sqrt{f'(t)}} + k_0^2\sigma^2\left(1 - \frac{2}{\pi}\right)\right]\Lambda_t dt, \tag{13}$$

with $\Lambda_t = ES_t^4\widehat{C}_{xx}^2(t, S_t)$. *Explicitly,*

$$\Lambda_t = \frac{1}{2\pi\rho_t}\frac{K^2}{\sqrt{2\sigma^2 t + \rho_t^2}}\exp\left\{-\frac{\left(\ln\frac{S_0}{K} - \frac{1}{2}\sigma^2 t - \frac{1}{2}\rho_t^2\right)^2}{2\sigma^2 t + \rho_t^2}\right\}. \tag{14}$$

The case $f(t) = t$ corresponds to the uniform grid and $A_1(f) = A_1$.

The above result makes plausible the conjecture that the normalized difference $n^{1/2}(V_1^n - V_1)$ converges in law. Indeed, this is the case, see [4].

1.2 Comments on the Grannan–Swindle Paper

The Leland method based on the Black–Scholes formula is amongst a few practical recipes how to price options under transaction costs. It has an advantage to rely upon well-known and well-understood formulae from the theory of frictionless markets. The method gave rise to a variety of other schemes. Of course, the precision of the resulting approximate hedging is an important issue, see [5], [2], [8], [9] and a survey [10] for related development.

The idea to parameterize the non-uniform grids by increasing functions and consider the family of strategies with the enlarged volatilities given by (10) is due to Grannan and Swindle, [3]. The mentioned paper claims that the asymptotics (12) holds for more general option with the pay-off of the form $G(S_1)$. In such a case the function $\widehat{C}(t,x)$ is the solution of the Cauchy problem

$$\widehat{C}_t(t,x) + \frac{1}{2}\widehat{\sigma}_t^2 x^2 \widehat{C}_{xx}(t,x) = 0, \qquad \widehat{C}(1,x) = G(x).$$

To our opinion, the formulations and arguments given in [3] are not satisfactory. In particular, the hypothesis that for any nonnegative integers m, n, p

$$\|\widehat{C}\|_{m,n,p} = \sup_{x>0,\, t\in[0,1]} \left[x^m \frac{\partial^{n+p}\widehat{C}(t,x)}{\partial x^n \partial t^p} \right] < \infty$$

is not fulfilled for the call-option with $G(x) = (x - K)^+$ (even for the uniform grid): explicit formulae show that derivatives of $\widehat{C}(t,x)$ have singularities at the point $(1, K)$. So, the mathematical results of the original paper [3] do not cover practically interesting cases. Nevertheless, the formula for $A_1(f)$ is used in numerical analysis of the approximate hedging error of call-options. Note also that the authors of [3] do not care about the eventual divergence of the integral (13) due to singularities of $1/f'$ which are not excluded by their assumptions. Neglecting the singularities may lead to an erroneous answer (recall the unfortunate error in Leland's paper corrected in [5]). That is why we are looking here for a rigorous proof to built a platform for further studies. The asymptotic analysis happens to be more involved comparatively with the arguments in [3] and we restrict ourselves to the case of the classical call-option.

The paper [3] contains another interesting idea: to minimize the functional $A_1(f)$ with respect to the scale f in a hope to improve the performance of the strategy by an appropriate choice of the revision dates[1]. We alert the reader that the reduction to a classical variational problem is not correct as well as the derived Euler–Lagrange equation. That is why the whole paper [3] can be considered only as one giving useful heuristics but leaving open mathematical problems of practical importance.

2. Proof of Theorem 1.2

2.1 Preparatory Manipulations

First of all, we represent the deviation of the approximating portfolio from the pay-off in an integral form which is instructive how to proceed further.

Lemma 2.1. *We have the representation* $V_1^n - V_1 = F_1^n + F_2^n$ *where*

$$F_1^n = \sigma \sum_{i=1}^{n} \int_{t_{i-1}}^{t_i} (\widehat{C}_x(t_{i-1}, S_{t_{i-1}}) - \widehat{C}_x(t, S_t)) S_t dW_t,$$

$$F_2^n = k_0 \sqrt{\frac{2}{\pi}} \sigma \int_0^1 S_t^2 \widehat{C}_{xx}(t, S_t) \sqrt{f'(t)} dt - \frac{k_0}{\sqrt{n}} \sum_{i=1}^{n-1} |\widehat{C}_x(t_i, S_{t_i}) - \widehat{C}_x(t_{i-1}, S_{t_{i-1}})| S_{t_i}.$$

[1] Even in the frictionless case the choice of an optimal scale to minimize the hedging error is an important and nontrivial problem, especially, for irregular pay-off functions, see, e.g., [1] and references wherein.

Proof. Applying the Ito formula and taking into account that the functions $C(t,x)$ and $\widehat{C}(t,x)$ satisfy the Cauchy problems with the same terminal condition, we get:

$$\begin{aligned}\widehat{C}(0,S_0) - C(0,S_0) &= C(1,S_1) - C(0,S_0) - [\widehat{C}(1,S_1) - \widehat{C}(0,S_0)] \\ &= \int_0^1 C_x(t,S_t)dS_t - \int_0^1 \widehat{C}_x(t,S_t)dS_t \\ &+ \int_0^1 \frac{1}{2}(\widehat{\sigma}_t^2 - \sigma^2)S_t^2\widehat{C}_{xx}(t,S_t)dt.\end{aligned}$$

Taking the difference of (4) and (5) with $t = 1$ and using the above formula, we obtain the required representation. □

Note that the sum in the expression for F_2^n does not include the term with $i = n$. Having in mind singularities of derivatives at the maturity, it is convenient to isolate the last summands in other sums and treat them separately.

A short inspection of the above formulae using the well-known helpful heuristics $\Delta S_t \approx \sigma S_t \Delta W_t \approx \sigma S_t \sqrt{dt}$ reveals that the main contributions in the first order Taylor approximations of increments originate from the derivatives in x. This consideration allows us to specify the principal terms of a particularly transparent structure. Namely, we put

$$\begin{aligned}P_1^n &:= \sum_{i=1}^{n-1} \sigma\widehat{C}_{xx}(t_{i-1},S_{t_{i-1}})S_{t_{i-1}}^2 \int_{t_{i-1}}^{t_i} (1 - S_t/S_{t_{i-1}})S_t/S_{t_{i-1}}dW_t, \\ P_2^n &:= k_0 \sum_{i=1}^{n-1} \widehat{C}_{xx}(t_{i-1},S_{t_{i-1}})S_{t_{i-1}}^2\left[\sigma\sqrt{2/\pi}\sqrt{f'(t_{i-1})}\Delta t_i - |S_{t_i}/S_{t_{i-1}} - 1|/\sqrt{n}\right].\end{aligned}$$

To establish Theorem 1.2 we check that $nE(P_1^n + P_2^n)^2 \to A_1(f)$ as $n \to \infty$ and the residual terms $R_i^n := F_i^n - P_i^n$ are negligible, i.e. $nE(R_i^n)^2 = o(1)$.

The first residual term is of the following form:

$$R_1^n = (R_{1n}^n - R_{1t}^n - (1/2)\tilde{R}_1^n)\sigma, \tag{15}$$

where

$$\begin{aligned}R_{1n}^n &= \int_{t_{n-1}}^1 (\widehat{C}_x(t_{n-1},S_{t_{n-1}}) - \widehat{C}_x(t,S_t))S_t dW_t, \\ R_{1t}^n &= \sum_{i=1}^{n-1} \widehat{C}_{xt}(t_{i-1},S_{t_{i-1}}) \int_{t_{i-1}}^{t_i} (t - t_{i-1})S_t dW_t, \\ \tilde{R}_1^n &= \sum_{i=1}^{n-1} \int_{t_{i-1}}^{t_i} \tilde{U}_t^i dW_t,\end{aligned}$$

where

$$\tilde{U}_t^i = \widehat{C}_{xxx}(\tilde{t}_{i-1}, \tilde{S}_{t_{i-1}})(S_t - S_{t_{i-1}})^2 S_t + \widehat{C}_{xtt}(\tilde{t}_{i-1}, \tilde{S}_{t_{i-1}})(t - t_{i-1})^2 S_t$$
$$+2\widehat{C}_{xxt}(\tilde{t}_{i-1}, \tilde{S}_{t_{i-1}})(t - t_{i-1})(S_t - S_{t_{i-1}})S_t,$$

$\tilde{t}_{i-1}$ and $\tilde{S}_{t_{i-1}}$ are random variables with values in the intervals $[t_{i-1}, t_i]$ and $[S_{t_{i-1}}, S_t]$, respectively. The structure of the above representation of R_1^n is clear: the term R_{1n}^n corresponds to the n-th revision interval (it will be treated separately because of singularities at the left extremity of the time interval), the term R_{1t}^n involving the first derivatives of $\widehat{C}_x$ in t at points $(t_{i-1}, S_{t_{i-1}})$ comes from the Taylor formula and the "tilde" term is due to the remainder of latter.

It is important to note that the integrals involving in the definition of P_1^n depend only on the increments of the Wiener process on the intervals $[t_{i-1}, t_i]$ and, therefore, are independent on the σ-algebras $\mathcal{F}_{t_{i-1}}$. This helps to calculate the expectation of the squared sum: according to Lemma 2.2 below it is the sum of expectations of the squared terms. We define P_2^n in a way to enjoy the same property. The second residual term includes the term R_{2n}^n corresponding to the last revision interval; the term R_{21}^n represents the approximation error arising from replacement of the integral by the Riemann sum; the remaining part of the residual we split in a natural way into summands R_{22}^n and R_{23}^n. After these explanations we write the second residual term as follows:

$$R_2^n = (R_{2n}^n + R_{21}^n + R_{22}^n + R_{23}^n + R_{24}^n)k_0, \tag{16}$$

with

$$R_{2n}^n = \sqrt{\frac{2}{\pi}}\sigma \int_{t_{n-1}}^1 S_t^2 \widehat{C}_{xx}(t, S_t)\sqrt{f'(t)}dt,$$

$$R_{21}^n = \sqrt{\frac{2}{\pi}}\sigma \sum_{i=1}^{n-1} \int_{t_{i-1}}^{t_i} (S_t^2 \widehat{C}_{xx}(t, S_t)\sqrt{f'(t)} - S_{t_{i-1}}^2 \widehat{C}_{xx}(t_{i-1}, S_{t_{i-1}})\sqrt{f'(t_{i-1})})dt,$$

$$R_{22}^n = \frac{1}{\sqrt{n}} \sum_{i=1}^{n-1} \widehat{C}_{xx}(t_{i-1}, S_{t_{i-1}})|S_{t_{i-1}} - S_{t_i}|(S_{t_{i-1}} - S_{t_i}),$$

$$R_{23}^n = \frac{1}{\sqrt{n}} \sum_{i=1}^{n-1} [...]_i (S_{t_i} - S_{t_{i-1}}),$$

$$R_{24}^n = \frac{1}{\sqrt{n}} \sum_{i=1}^{n-1} [...]_i S_{t_{i-1}},$$

where

$$[...]_i = \widehat{C}_{xx}(t_{i-1}, S_{t_{i-1}})|S_{t_i} - S_{t_{i-1}}| - |\widehat{C}_x(t_i, S_{t_i}) - \widehat{C}_x(t_{i-1}, S_{t_{i-1}})|. \tag{17}$$

2.2 Tools

In our computations we shall use frequently the following two assertions. The first one is a standard fact on square integrable martingales in discrete time.

Lemma 2.2. *Let $M = (M_i)$ be a square-integrable martingale with respect to a filtration $(\mathcal{G}_i)$, $i = 0, ..., k$, and let $X = (X_i)$ be a predictable process with $EX^2 \cdot \langle M\rangle_k < \infty$. Then*

$$E(X \cdot M_k)^2 = EX^2 \cdot \langle M\rangle_k = \sum_{i=1}^{k} EX_i^2(\Delta M_i)^2,$$

where, as usual, $\Delta\langle M\rangle_i := E((\Delta M_i)^2|\mathcal{G}_{i-1})$,

$$X \cdot M_k := \sum_{i=1}^{k} X_i \Delta M_i, \qquad X^2 \cdot \langle M\rangle_k := \sum_{i=1}^{k} X_i^2 \langle M\rangle_i.$$

Lemma 2.3. *Suppose that $g', f' \in C([0,1])$. Let $p > 0$ and $a \geq 0$. Then*

$$\sum_{i=1}^{n-1} \frac{(\Delta t_i)^{p+a}}{(1-t_i)^p} = \begin{cases} O(n^{1-p-a}), & p < 1, \\ O(n^{-a} \ln n), & p = 1, \\ O(n^{-a}), & p > 1. \end{cases}$$

If $g(t) = 1 - (1-t)^\beta$, $\beta \geq 1$, then

$$\sum_{i=1}^{n-1} \frac{(\Delta t_i)^{p+a}}{(1-t_i)^p} = \begin{cases} O(n^{1-p-a}), & p < 1 + a(\beta - 1), \\ O(n^{-a\beta} \ln n), & p = 1 + a(\beta - 1), \\ O(n^{-a}), & p > 1 + a(\beta - 1). \end{cases}$$

Proof. We consider first the case where $g', f' \in C([0,1])$, i.e. g' is not only bounded but also bounded away from zero. By the finite increments formula $\Delta t_i = g'(x_i)n^{-1}$ where $x_i \in [(i-1)/n, i/n]$ and, hence, $\Delta t_i \leq \text{const}\, n^{-1}$. Applying again the finite increments formula and taking into account that $\min g'(t) > 0$, it is easy to check that there is a constant c such that

$$\frac{1-t_{i-1}}{1-t_i} \leq c, \qquad 1 \leq i \leq n-1.$$

Thus,

$$\sum_{i=1}^{n-1} \frac{\Delta t_i}{(1-t_i)^p} \leq c \sum_{i=1}^{n-1} \frac{\Delta t_i}{(1-t_{i-1})^p} \leq c \int_0^{t_{n-1}} \frac{dt}{(1-t)^p}.$$

Since

$$n^{-1} \min g'(t) \leq 1 - g(1 - 1/n) \leq n^{-1} \max g'(t),$$

the asymptotic of the last integral is $O(1)$, if $p < 1$ (the integral converges), $O(\ln n)$, if $p = 1$, and $O(n^{p-1})$, if $p > 1$,. This implies the claimed property.

In the second case where $g(t) = 1 - (1-t)^\beta, \beta \geq 1$, we have

$$\sum_{i=1}^{n-1} \frac{(\Delta t_i)^{p+a}}{(1-t_i)^p} = \frac{\beta^{p+a}}{n^{p-1+a}} \sum_{i=1}^{n-1} \frac{(1-x_i)^{(\beta-1)(p+a)}}{(1-i/n)^{\beta p}} \frac{1}{n}.$$

The sum in the right-hand side is dominated, up to a multiplicative constant, by

$$\sum_{i=1}^{n-1} \frac{1}{(1-(i-1)/n)^{p+a-\beta a}} \frac{1}{n} \leq \int_0^{1-1/n} \frac{dt}{(1-t)^{p+a-\beta a}}.$$

Using the explicit formulae for the integral we infer that the required property holds whatever are the parameters $p > 0$, $a \geq 0$, and $\beta \geq 1$. □

2.3 Explicit Formulae and Useful Bounds

We consider the function $\widehat{C}(t,x)$ corresponding to the "artificial", in general, time-varying volatility. This function, solving the Cauchy problem (11), is given by the formula

$$\widehat{C}(t,x) = x\Phi(\widehat{d}(t,x)) - K\Phi(\widehat{d}(t,x) - \rho_t), \qquad t < 1, \tag{18}$$

where

$$\widehat{d}(t,x) := \frac{1}{\rho_t} \ln \frac{x}{K} + \frac{1}{2}\rho_t \tag{19}$$

and $\rho_t > 0$,

$$\rho_t^2 := \int_t^1 \widehat{\sigma}_s^2 ds = \int_t^1 (\sigma^2 + \sigma k_0 \sqrt{8/\pi} \sqrt{f'(s)}) ds. \tag{20}$$

It is easy to verify that

$$\widehat{C}_x(t,x) = \Phi(\widehat{d}(t,x)),$$
$$\widehat{C}_{xx}(t,x) = \frac{1}{x\rho_t} \varphi(\widehat{d}(t,x)).$$

The first and the second derivatives of $\widehat{d}(t,x)$ in t are, respectively,

$$\widehat{d}_t(t,x) = \frac{\widehat{\sigma}_t^2}{2\rho_t^3} \ln \frac{x}{K} - \frac{\widehat{\sigma}_t^2}{4\rho_t} = \frac{\widehat{\sigma}_t^2}{2\rho_t^2} (\widehat{d}(t,x) - \rho_t),$$
$$\widehat{d}_{tt}(t,x) = \frac{\sigma k_0 \sqrt{2/\pi} f''(t)}{2\rho_t^2 \sqrt{f'(t)}} (\widehat{d}(t,x) - \rho_t) + \frac{3\widehat{\sigma}_t^4}{4\rho_t^4} (\widehat{d}(t,x) - \rho_t) + \frac{\widehat{\sigma}_t^4}{4\rho_t^3}.$$

For the analysis of residual terms we need also the following derivatives:

$$\widehat{C}_{xt}(t,x) = \frac{\widehat{\sigma}_t^2}{2\rho_t^2}(\widehat{d}(t,x) - \rho_t)\varphi(\widehat{d}(t,x)),$$

$$\widehat{C}_{xxx}(t,x) = -\frac{1}{x^2\rho_t^2}(\widehat{d}(t,x) + \rho_t)\varphi(\widehat{d}(t,x)),$$

$$\widehat{C}_{xtt}(t,x) = \left[\widehat{d}_{tt}(t,x) - \frac{\widehat{\sigma}_t^4}{4\rho_t^4}(\widehat{d}(t,x) - \rho_t)^2\widehat{d}(t,x)\right]\varphi(\widehat{d}(t,x)),$$

$$\widehat{C}_{xxt}(t,x) = \frac{1}{x}\left[\frac{\widehat{\sigma}_t^2}{2\rho_t^3} + \frac{\widehat{\sigma}_t^2}{2\rho_t^3}(\widehat{d}(t,x) - \rho_t)\widehat{d}(t,x)\right]\varphi(\widehat{d}(t,x)),$$

$$\widehat{C}_{xxxx}(t,x) = \frac{1}{x^3\rho_t^3}\left[2\rho_t(\widehat{d}(t,x) + \rho_t) - 1 + (\widehat{d}(t,x) + \rho_t)\widehat{d}(t,x)\right]\varphi(\widehat{d}(t,x)).$$

Note that under any of our assumptions on the scale transformation the ratio $\widehat{\sigma}_t^2/\rho_t^2$ has a singularity $c(1-t)^{-1}$ as $t \to 1$. Since for every $p \geq 0$ the function $|y|^p\varphi(y)$ is bounded, we obtain from the above formulae the following estimates:

$$|\widehat{C}_{xt}(t,x)| \leq \kappa\frac{1}{1-t}, \tag{21}$$

$$|\widehat{C}_{xxx}(t,x)| \leq \kappa\frac{1}{x^2(1-t)}, \tag{22}$$

$$|\widehat{C}_{xtt}(t,x)| \leq \kappa\frac{1}{(1-t)^2}, \tag{23}$$

$$|\widehat{C}_{xxt}(t,x)| \leq \kappa\frac{1}{x}\frac{1}{(1-t)^{3/2}}, \tag{24}$$

$$|\widehat{C}_{xxxx}(t,x)| \leq \kappa\frac{1}{x^3(1-t)^{3/2}}. \tag{25}$$

Now we obtain a formula which gives, in particular, an expression for Λ_t.

Let $\xi \in \mathcal{N}(0,1)$ and let $a \neq 0$, b, c be arbitrary constants. Then

$$Ee^{c\xi}e^{-(a\xi+b)^2} = \frac{1}{\sqrt{2a^2+1}}\exp\left\{-\frac{\tilde{b}^2}{2a^2+1} + \tilde{b}^2 - b^2\right\}. \tag{26}$$

where $\tilde{b} := b - c/(2a)$.

The distribution of the random variable $2\pi S_t^p\widehat{C}_{xx}^2(t,S_t)$ is the same as of

$$S_0^{p-2}e^{-\frac{1}{2}(p-2)\sigma^2 t}\rho_t^{-2}e^{c_t\xi}e^{-(a_t\xi+b_t)^2}$$

where $c_t = (p-2)\sigma t^{1/2}$, $a_t = \frac{1}{\rho_t}\sigma t^{1/2}$,

$$b_t = \frac{1}{\rho_t}\left(\ln\frac{S_0}{K} - \frac{1}{2}\sigma^2 t\right) + \frac{1}{2}\rho_t, \qquad \tilde{b}_t = b_t - \frac{1}{2}(p-2)\rho_t.$$

Since

$$\tilde{b}_t^2 - b_t^2 = -(p-2)\left[\left(\ln\frac{S_0}{K} - \frac{1}{2}\sigma^2 t\right) + \rho_t^2 - \frac{1}{4}p\rho_t^2\right],$$

we obtain from above that

$$ES_t^p\widehat{C}_{xx}^2(t,S_t) = \frac{1}{2\pi\rho_t}\frac{K^{p-2}}{\sqrt{2\sigma^2 t + \rho_t^2}}e^{-B_t}, \tag{27}$$

where

$$B_t := \frac{\left(\ln\frac{S_0}{K} - \frac{1}{2}\sigma^2 t - \frac{1}{2}(p-3)\rho_t^2\right)^2}{2\sigma^2 t + \rho_t^2} - \frac{(p-2)(p-4)}{4}\rho_t^2. \tag{28}$$

In particular, with $p = 4$, we have the following formula:

$$\Lambda_t = \frac{1}{2\pi\rho_t}\frac{K^2}{\sqrt{2\sigma^2 t + \rho_t^2}}\exp\left\{-\frac{\left(\ln\frac{S_0}{K} - \frac{1}{2}\sigma^2 t - \frac{1}{2}\rho_t^2\right)^2}{2\sigma^2 t + \rho_t^2}\right\}. \tag{29}$$

It is easily seen that Λ_t is a continuous function on $[0,1[$ tending to infinity as $t \to 1$. The singularity at infinity is integrable since

$$\frac{1}{\kappa}\frac{1}{(1-t)^{1/4}} \le \Lambda_t \le \kappa\frac{1}{(1-t)^{1/2}} \tag{30}$$

for some constant $\kappa > 0$ (of course, under Assumption 1 we have the lower bound of the same order as the upper one).

It is worth to notice that the upper bound above is better than one could get using the straightforward estimate $\widehat{C}_{xx}^2(t,x) \le \kappa/(x^2\rho_t^2)$.

Sharper bounds for the expectations will be of frequent use in our analysis. To get them we observe that for $p \in \mathbf{R}$, $m \ge 0$, $r > 0$, and $r' \in]0,r[$ we have the bound

$$ES_t^p\widehat{d}^m(t,S_t)\varphi(r\widehat{d}(t,S_t)) \le \kappa ES_t^p\varphi(r'\widehat{d}(t,S_t)).$$

Exploiting again the identity (26) to estimate the right-hand side we get the inequality

$$ES_t^p\widehat{d}^m(t,S_t)\varphi(r\widehat{d}(t,S_t)) \le \kappa\rho_t.$$

Using the expressions for the derivatives of $\widehat{C}$ we obtain the following bounds:

$$ES_t^p \widehat{C}_{xt}^{2m}(t, S_t) \le \kappa \frac{1}{(1-t)^{2m-1/2}}, \tag{31}$$

$$ES_t^p \widehat{C}_{xxx}^{2m}(t, S_t) \le \kappa \frac{1}{(1-t)^{2m-1/2}}, \tag{32}$$

$$ES_t^p \widehat{C}_{xxt}^{2m}(t, S_t) \le \kappa \frac{1}{(1-t)^{3m-1/2}}, \tag{33}$$

$$ES_t^p \widehat{C}_{xxxx}^{2m}(t, S_t) \le \kappa \frac{1}{(1-t)^{3m-1/2}}, \tag{34}$$

where the constant κ depends on p and m.

2.4 Analysis of the Principal Terms

Let us check that $nE(P_1^n + P_2^n) \to A_1(f)$ as $n \to \infty$. To this aim we put

$$\tilde{P}_1^n := \frac{1}{2}\sigma^2 \sum_{i=1}^{n-1} \widehat{C}_{xx}(t_{i-1}, S_{t_{i-1}}) S_{t_{i-1}}^2 [\Delta t_i - (\Delta W_{t_i})^2],$$

$$\tilde{P}_2^n := k_0 \sigma n^{-1/2} \sum_{i=1}^{n-1} \widehat{C}_{xx}(t_{i-1}, S_{t_{i-1}}) S_{t_{i-1}}^2 \left[\sqrt{2/\pi}\sqrt{\Delta t_i} - |\Delta W_{t_i}| \right],$$

where, as usual, $\Delta t_i := t_i - t_{i-1}$ and $\Delta W_{t_i} := W_{t_i} - W_{t_{i-1}}$. It is sufficient to verify that $nE(\tilde{P}_1^n + \tilde{P}_2^n)^2 \to A_1(f)$ while $nE(P_j^n - \tilde{P}_j^n)^2 \to 0$, $j = 1, 2$.

Recall that $E(\xi^2 - 1)^2 = 2$ and $E|\xi|^3 = 2E|\xi| = 2\sqrt{2/\pi}$ for $\xi \in \mathcal{N}(0, 1)$. Using Lemma 2.2 we obtain the representation

$$nE(\tilde{P}_1^n + \tilde{P}_2^n)^2 = \frac{\sigma^4}{2} n \sum_{i=1}^{n-1} \Lambda_{t_{i-1}} (\Delta t_i)^2 + k_0 \sigma^3 \sqrt{\frac{2}{\pi}} n^{1/2} \sum_{i=1}^{n-1} \Lambda_{t_{i-1}} (\Delta t_i)^{3/2}$$
$$+ k_0^2 \sigma^2 \Big(1 - \frac{2}{\pi}\Big) \sum_{i=1}^{n-1} \Lambda_{t_{i-1}} \Delta t_i.$$

By the finite increments formula $\Delta t_i = g(i/n) - g((i-1)/n) = g'(x_i)/n$ where $x_i \in [(i-1)/n, i/n]$. We substitute this expression into the sums above. Let us introduce the function F_n (depending on p) by the formula

$$F_n(t) := \sum_{i=1}^{n-1} \Lambda_{g((i-1)/n)} [g'(x_i)]^p I_{[(i-1)/n, i/n[}(t).$$

For $p \ge 1$ we have:

$$\sum_{i=1}^{n-1} \Lambda_{g((i-1)/n)} [g'(x_i)]^p \frac{1}{n} = \int_0^1 F_n(t)dt \to \int_0^1 \Lambda_{g(t)} [g'(t)]^p dt.$$

The needed uniform integrability of the sequence $\{F_n\}$ with respect to the Lebesgue measure follows from the de la Vallée-Poussin criterion because the estimate $\Lambda_t \le \kappa(1-t)^{-1/2}$ and the boundedness of g' imply that

$$\int_0^1 F_n^{3/2}(t)dt \le \text{const} \int_0^1 \frac{dg(t)}{(1-g(t))^{3/4}} = \text{const} \int_0^1 \frac{ds}{(1-s)^{3/4}} < \infty.$$

By the change of variable, taking into account that $g'(t) = 1/f'(g(t))$, we transform the limiting integral into the form used in the formulations of the theorem:

$$\int_0^1 \Lambda_{g(t)}[g'(t)]^p dt = \int_0^1 \Lambda_{g(t)}[g'(t)]^{p-1} dg(t) = \int_0^1 \Lambda_t [f'(t)]^{1-p} dt.$$

The first claimed property on the convergence to $A_1(f)$ is verified.

Using again Lemma 2.2 we get that

$$E(P_1^n - \tilde{P}_1^n)^2 = \sigma^2 \sum_{i=1}^{n-1} \Lambda_{t_{i-1}} \int_{t_{i-1}}^{t_i} E\Big[\Big(\frac{S_t}{S_{t_{i-1}}} - 1\Big)\frac{S_t}{S_{t_{i-1}}} - \sigma(W_t - W_{t_{i-1}})\Big]^2 dt.$$

It is a simple exercises to check that

$$E((e^{\sigma t^{1/2}\xi - \frac{1}{2}\sigma^2 t} - 1)e^{\sigma t^{1/2}\xi - \frac{1}{2}\sigma^2 t} - \sigma t^{1/2}\xi)^2 = O(t^2), \quad t \to 0.$$

Therefore,

$$nE(P_1^n - \tilde{P}_1^n)^2 \le \text{const}\, n \sum_{i=1}^{n-1} \Lambda_{t_{i-1}} (\Delta t_i)^3 \to 0, \quad n \to 0.$$

The sum P_2^n is not centered and, therefore, Lemma 2.2 cannot be directly applied. Let us check that under our assumptions the bias is negligible. Indeed, put

$$P_2^{n\prime} = k_0 n^{-1/2} \sum_{i=1}^{n-1} \widehat{C}_{xx}(t_{i-1}, S_{t_{i-1}}) S_{t_{i-1}}^2 \Big[E|S_{t_i}/S_{t_{i-1}} - 1| - |S_{t_i}/S_{t_{i-1}} - 1|\Big].$$

We have:

$$n^{1/2} \|P_2^n - P_2^{n\prime}\|_{L^2} \le k_0 \sum_{i=1}^{n-1} \Lambda_{t_{i-1}}^{1/2} B_i,$$

where

$$B_i := \Big|\sigma\sqrt{2/\pi}\sqrt{nf'(t_{i-1})}\Delta t_i - E|S_{t_i}/S_{t_{i-1}} - 1|\Big|.$$

Using the Taylor formula it is easy to verify that for $u > 0$

$$E|e^{u\xi - \frac{1}{2}u^2} - 1| = 2[\Phi(u/2) - \Phi(-u/2)] = \sqrt{2/\pi}u + O(u^3), \quad u \to 0,$$

It follows that

$$B_i = \sigma \sqrt{2/\pi}(\Delta t_i)^{1/2} \Big| \sqrt{n f'(t_{i-1})\Delta t_i} - 1| + O((\Delta t_i)^{3/2}).$$

By the Taylor formula

$$\Delta t_i = g(i/n) - g((i-1)/n) = g'((i-1)/n)\frac{1}{n} + \frac{1}{2}g''(y_i)\frac{1}{n^2},$$

where the point $y_i \in [(i-1)/n, i/n]$. Since the function f is the inverse of g we have $f'(t_{i-1}) = 1/g'((i-1)/n)$. Using these identities and the elementary inequality $|\sqrt{1+a} - 1| \le |a|$ for $a \ge -1$ we obtain that

$$B_i \le \text{const}\, \frac{|g''(y_i)|}{g'((i-1)/n)}(\Delta t_i)^{1/2}\frac{1}{n} + O((\Delta t_i)^{3/2}).$$

Fix $\varepsilon \in]0, 1/4[$. Substituting the finite increments formula $\Delta t_i = g'(x_i)/n$ with an intermediate point x_i in $[(i-1)/n, i/n]$, we infer that

$$B_i \le \text{const}\, a_n \frac{g'(x_i)}{[1 - g((i-1)/n)]^{3/4-\varepsilon}}\frac{1}{n} + O((\Delta t_i)^{3/2}).$$

where

$$a_n = \frac{1}{n^{1/2}} \sup_{i \le n-1} \sup_{x_i, y_i} \frac{|g''(y_i)|[1 - g((i-1)/n)]^{3/4-\varepsilon}}{g'((i-1)/n)(g'(x_i))^{1/2}}.$$

Recall that

$$\sum_{i=1}^{n-1} \frac{g'(x_i)}{[1 - g((i-1)/n)]^{1-\varepsilon}}\frac{1}{n} \to \int_0^1 \frac{dg(t)}{[1 - g(t)]^{1-\varepsilon}} = \int_0^1 \frac{dt}{(1-t)^{1-\varepsilon}} < \infty$$

and $a_n \to 0$ under each of our assumptions. These observations lead to the conclusion that

$$\sum_{i=1}^{n-1} \Lambda_{t_{i-1}}^{1/2} B_i \to 0.$$

Thus, we have the convergence $nE(P_2^n - P_2^{n\prime})^2 \to 0$.

Noticing that

$$E(|e^{u\xi - \frac{1}{2}u^2} - 1| - u|\xi|)^2 = O(u^4), \qquad u \to 0,$$

we infer that

$$E\Big[\Big(E|S_{t_i}/S_{t_{i-1}} - 1| - |S_{t_i}/S_{t_{i-1}} - 1|\Big) - \sigma\Big(E|\Delta W_{t_i}| - |\Delta W_{t_i}|\Big)\Big]^2 = O((\Delta t_i)^2).$$

Using again Lemma 2.2 we get that

$$nE(P_2^{n\prime} - \tilde{P}_2^n)^2 \le \text{const} \sum_{i=1}^{n-1} \Lambda_{t_{i-1}}(\Delta t_i)^2 \to 0, \quad n \to \infty.$$

It follows that $nE(P_2^n - \tilde{P}_2^n)^2 \to 0$ and the proof of the second claimed property is completed.

2.5 Analysis of the Residual R_1^n

1. Calculating the expectation of the squared stochastic integral we obtain that

$$E(R_{1n}^n)^2 = \int_{t_{n-1}}^{1} E(\widehat{C}_x(t_{n-1}, S_{t_{n-1}}) - \widehat{C}_x(t, S_t))^2 S_t^2 dt \le \kappa_n(1 - t_{n-1}),$$

where κ_n is the supremum of the integrand over $[t_{n-1}, 1]$. Note that the process $\widehat{d}(t, S_t)$ diverges a.s. when $t \to 1$. On the set where it diverges to $-\infty$, we have $\widehat{C}_x(t, S_t) = \Phi(\widehat{d}(t, S_t)) \to 0$ while $\widehat{C}_x(t, S_t) \to 1$ a.s. on the complement of this set. It follows that $\kappa_n \to 0$. Since $1 - t_{n-1} \le \kappa n^{-1}$ (due to the boundedness of g'), we conclude that $nE(R_{1n}^n)^2 \to 0$.

2. Let us consider the term

$$R_{1t}^n = \sum_{i=1}^{n-1} \widehat{C}_{xt}(t_{i-1}, S_{t_{i-1}}) \int_{t_{i-1}}^{t_i} (t - t_{i-1}) S_t dW_t.$$

According to (31)

$$E\widehat{C}_{xt}^2(t, S_t) S_t^2 \le \kappa \frac{1}{(1-t)^{3/2}}.$$

Therefore,

$$\begin{aligned} E(R_{1t}^n)^2 &= \sum_{i=1}^{n-1} E\widehat{C}_{xt}^2(t_{i-1}, S_{t_{i-1}}) S_{t_{i-1}}^2 \int_{t_{i-1}}^{t_i} (t - t_{i-1})^2 E(S_t / S_{t_{i-1}})^2 dt \\ &\le \text{const} \sum_{i=1}^{n-1} \frac{(\Delta t_i)^3}{(1 - t_{i-1})^{3/2}} = O(n^{-3/2}), \qquad n \to \infty, \end{aligned}$$

in virtue of Lemma 2.3. Hence, $nE(R_{1t}^n)^2 \to 0$.

3. Now we estimate the expectation $E(\tilde{R}_1^n)^2$ corresponding to the terms arising from the residual in the Taylor formula for $\widehat{C}_x$. We have:

$$E(\tilde{R}_1^n)^2 = \sum_{i=1}^{n-1} \int_{t_{i-1}}^{t_i} E(\tilde{U}_t^i)^2 dt.$$

Since $(a+b+c)^2 \le 3(a^2+b^2+c^2)$, it is sufficient to check that each of the following sums converge to zero as $o(n^{-1})$:

$$\Sigma_1^n = \sum_{i=1}^{n-1} \int_{t_{i-1}}^{t_i} E\widehat{C}_{xxx}^2(\tilde{t}_{i-1}, \tilde{S}_{t_{i-1}})(S_t - S_{t_{i-1}})^4 S_t^2 dt,$$

$$\Sigma_2^n := \sum_{i=1}^{n-1} \int_{t_{i-1}}^{t_i} E\widehat{C}_{xtt}^2(\tilde{t}_{i-1}, \tilde{S}_{t_{i-1}})(t - t_{i-1})^4 S_t^2 dt,$$

$$\Sigma_3^n := \sum_{i=1}^{n-1} \int_{t_{i-1}}^{t_i} E\widehat{C}_{xxt}^2(\tilde{t}_{i-1}, \tilde{S}_{t_{i-1}})(t - t_{i-1})^2 (S_t - S_{t_{i-1}})^2 S_t^2 dt.$$

Taking into account that $y\varphi(y)$ is bounded, $\widehat{d}(t,x)\varphi(\widehat{d}(t,x)) \to 0$ as $t \to 1$ whatever is $x \neq K$, and $S_1 \neq K$ (a.s), we deduce from the explicit formula for $\widehat{C}_{xxx}$ that for any $\varepsilon > 0$, $m \geq 1$, there exists $a \in]0,1[$ such that

$$E|\widehat{C}_{xxx}(\tilde{t}_{i-1}, \tilde{S}_{t_{i-1}})|^{2m} \leq \varepsilon \frac{1}{(1-t_i)^{2m}} \tag{35}$$

for every $t_{i-1} \geq a$. For $t_{i-1} < a$ the above expectation is bounded by a constant which does not on n.

Let $\xi \sim N(0,1)$ and let $b \in [0,1]$. Using the elementary bound

$$|e^{bx} - 1| \leq b(e^{|x|} - 1)$$

which follows from the Taylor expansion, we obtain, for $m \geq 1$, the estimate

$$E(e^{u\sigma\xi-(1/2)\sigma^2 u^2} - 1)^{2m} \leq \kappa u^{2m}$$

where the constant κ depends on m and σ. Applying the Cauchy–Schwarz inequality and this estimate we get that

$$E(S_t - S_{t_{i-1}})^{2m} S_t^p \leq \kappa (t - t_{i-1})^m.$$

Manipulating again with the Cauchy–Schwarz inequality we obtain with the help of the above bounds that

$$\Sigma_1^n \leq \kappa \sum_{t_{i-1}<a} (\Delta t_i)^3 + \kappa\varepsilon \sum_{i=1}^{n-1} \frac{(\Delta t_i)^3}{(1-t_i)^2}.$$

The first sum in the right-hand side is of order $O(n^{-2})$. According to Lemma 2.3 the second one is of order $O(n^{-1})$. Since $\varepsilon > 0$ is arbitrary, it follows that $\lim_n n\Sigma_1^n = 0$.

Similarly to the bound (35), we can establish that for any $\varepsilon > 0$ there is a threshold $a \in]0,1]$ such that for any $t_{i-1} \geq a$ the following inequalities hold:

$$E|\widehat{C}_{xxt}(\tilde{t}_{i-1}, \tilde{S}_{t_{i-1}})|^{2m} \leq \varepsilon \frac{1}{(1-t_i)^{3m}} \tag{36}$$

and

$$E|\widehat{C}_{xtt}(\tilde{t}_{i-1}, \tilde{S}_{t_{i-1}})|^{2m} \leq \varepsilon \frac{1}{(1-t_i)^{4m}}. \tag{37}$$

With these bounds we prove, making obvious changes in arguments, that $\lim_n n\Sigma_2^n = 0$ and $\lim_n n\Sigma_3^n = 0$.

2.6 Analysis of the Residual R_2^n

1. Noting that $\|S_t^2\widehat{C}_{xx}(t,S_t)\|_{L^2} = \Lambda_t^{1/2}$, we have:

$$\|R_{2n}^n)\|_{L^2} \le c\int_{t_{n-1}}^1 \Lambda_t^{1/2}\sqrt{f'(t)}dt \le c\left(\int_{t_{n-1}}^1 \Lambda_t dt\right)^{1/2}(1-f(t_{n-1}))^{1/2}$$

with $c = \sqrt{2/\pi}\sigma$. Since $f(t_{n-1}) = f(g((n-1)/n)) = 1 - 1/n$ and the function Λ is integrable, it follows that $nE(R_{2n}^n)^2 \to 0$.

2. The term R_{21}^n describes the error in approximation of an integral by Riemann sums. To analyze the approximation rate we need the following auxiliary result.

Lemma 2.4. *Let $X = (X_t)_{t\in[0,1]}$ be a process with*

$$dX_t = \mu_t dt + \vartheta_t dW_t, \qquad X_0 = 0,$$

where $\mu = (\mu_t)_{t\in[0,1]}$ and $\vartheta = (\vartheta_t)_{t\in[0,1]}$ are predictable processes such that

$$\int_0^1 (|\mu_t| + \vartheta_t^2)dt < \infty.$$

Let $X_t^n := \sum_{i=1}^n X_{t_{i-1}} I_{]t_{i-1},t_i]}(t)$. Then

$$E\left(\int_0^1 (X_t - X_t^n)dt\right)^2 \le 2\int_0^1 \sum_{i=1}^n (t_i-u)^2 I_{]t_{i-1},t_i]}(u)E\vartheta_u^2 du$$
$$+2\left(\int_0^1 \sum_{i=1}^n (t_i-u)I_{]t_{i-1},t_i]}(u)(E\mu_u^2)^{1/2}du\right)^2.$$

Proof. It is sufficient to work assuming that the right-hand side of the inequality is finite and consider separately the cases where one of the coefficients is zero. Let us start with the case where $\mu = 0$. Using the stochastic Fubini theorem, we have:

$$\int_{t_{i-1}}^{t_i}(X_t - X_{t_{i-1}})dt = \int_{t_{i-1}}^{t_i}\int_{t_{i-1}}^{t_i}\vartheta_u I_{]t_{i-1},t]}(u)dW_u dt = \int_{t_{i-1}}^{t_i}(t_i-u)\vartheta_u dW_u.$$

It follows that

$$E\left(\int_0^1 (X_t - X_t^n)dt\right)^2 = \int_0^1 \sum_{i=1}^n (t_i-u)^2 I_{]t_{i-1},t_i]}(u)E\vartheta_u^2 du.$$

In the case where $\vartheta = 0$ we have, this time by the ordinary Fubini theorem, that

$$\int_{t_{i-1}}^{t_i}(X_t - X_{t_{i-1}})dt = \int_{t_{i-1}}^{t_i}(t_i-u)\mu_u du$$

and this representation allows us to transform the left-hand side of the required inequality to the following form:

$$\int_0^1 \int_0^1 \sum_{i,j=1}^{n} (t_i - u)(t_j - v) I_{]t_{i-1},t_i]}(u) I_{]t_{j-1},t_j]}(v) E\mu_u\mu_v du dv.$$

Using the Cauchy–Schwarz inequality $E\mu_u\mu_v \le (E\mu_u^2)^{1/2}(E\mu_v^2)^{1/2}$ and once again the Fubini theorem we obtain the needed bound. □

Note that

$$E(R_{21}^n)^2 = E\left(\int_0^{t_{n-1}} (X_t - X_t^n)dt\right)^2$$

where $X_t := S_t^2 \widehat{C}_{xx}(t, S_t)\sqrt{f'(t)}$ has the coefficients

$$\begin{aligned}
\vartheta_t &= [2S_t\widehat{C}_{xx}(t,S_t) + S_t^2\widehat{C}_{xxx}(t,S_t)]\sqrt{f'(t)}\sigma S_t, \\
\mu_t &= \frac{1}{2}\left[2\widehat{C}_{xx}(t,S_t) + 4S_t\widehat{C}_{xxx}(t,S_t) + S_t^2\widehat{C}_{xxxx}(t,S_t)\right]\sqrt{f'(t)}\sigma^2 S_t^2 \\
&\quad + \frac{1}{2}S_t^2\widehat{C}_{xx}(t,S_t)\frac{f''(t)}{\sqrt{f'(t)}} + S_t^2\widehat{C}_{xxt}(t,S_t)\sqrt{f'(t)}.
\end{aligned}$$

In the case where g' is bounded away from zero (hence, f' is bounded), the estimates (27) and (32) imply that $E\vartheta_t^2 \le \kappa/(1-t)^{3/2}$. If also f'' is bounded, then the estimates (27) and (32) – (34) ensure that $E\mu_t^2 \le \kappa/(1-t)^{5/2}$.

Applying the previous lemma we have:

$$E(R_{21}^n)^2 \le \kappa \sum_{i=1}^{n-1} \frac{(\Delta t_i)^3}{(1-t_i)^{3/2}} + \kappa\left(\sum_{i=1}^{n-1} \frac{(\Delta t_i)^2}{(1-t_i)^{5/4}}\right)^2.$$

According to Lemma 2.3 the right-hand side is $O(n^{-3/2})$ as $n \to \infty$.

In the case where $g(t) = 1 - (1-t)^\beta$, $\beta > 1$, we obtain in the same way that $E\vartheta_t^2 \le \kappa/(1-t)^{5/2-1/\beta}$, $E\mu_t^2 \le \kappa/(1-t)^{7/2-1/\beta}$, and

$$E(R_{21}^n)^2 \le \kappa \sum_{i=1}^{n-1} \frac{(\Delta t_i)^3}{(1-t_i)^{5/2-1/\beta}} + \kappa\left(\sum_{i=1}^{n-1} \frac{(\Delta t_i)^2}{(1-t_i)^{7/4-1/(2\beta)}}\right)^2.$$

By Lemma 2.3 the first sum in the right-hand side can be of order $O(n^{-2})$, $O(n^{-2}\ln n)$, or $O(n^{-(\beta/2+1)})$, that is $o(n^{-1})$ as $n \to \infty$. The second sum can be $O(n^{-1})$, $O(n^{-1}\ln n)$, or $O(n^{-(\beta/4+1/2)})$, i.e. $o(n^{-1/2})$. In all cases $nE(R_{21}^n)^2 \to 0$.

3. The analysis of the term R_{22}^n is based on the first claim of Lemma 3.1 given in the section on asymptotics of Gaussian integrals.

We need to check that

$$E\left(\sum_{i=1}^{n-1}\widehat{C}_{xx}(t_{i-1},S_{t_{i-1}})S^2_{t_{i-1}}(S_{t_i}/S_{t_{i-1}}-1)^2\text{sign}\,(S_{t_i}/S_{t_{i-1}}-1)\right)^2$$

tends to zero.

For the expectation of the sum of squared terms we have:

$$\sum_{i=1}^{n-1}\Lambda_{t_{i-1}}E(S_{t_i}/S_{t_{i-1}}-1)^4=O(n^{-1}),$$

since

$$E(e^{u\xi-\frac{1}{2}u^2}-1)^4=O(u^4),\qquad u\to 0,$$

Λ_t is an integrable function on $[0,1]$, and $g'(t)$ is bounded.

Let us consider a "generic" cross term with indices $i<j$. It can be split in the product of two independent random variables. The expectation of the first one, $(S_{t_j}/S_{t_{j-1}}-1)^2\text{sign}\,(S_{t_j}/S_{t_{j-1}}-1)$, by virtue of Lemma 3.1 is dominated by $\kappa\Delta t_j^{3/2}$ where κ is a constant. The second one is the product of $\widehat{C}_{xx}(t_{i-1},S_{t_{i-1}})S^2_{t_{i-1}}(S_{t_i}/S_{t_{i-1}}-1)^2\text{sign}\,(S_{t_i}/S_{t_{i-1}}-1)$ and $\widehat{C}_{xx}(t_{j-1},S_{t_{j-1}})S^2_{t_{j-1}}$ and we dominate the absolute value of its expectation using the Cauchy–Schwarz inequality which gives the following bound:

$$\Lambda^{1/2}_{t_{i-1}}(E(S_{t_i}/S_{t_{i-1}}-1)^4)^{1/2}\Lambda^{1/2}_{t_{j-1}}\le\kappa\Lambda^{1/2}_{t_{i-1}}\Lambda^{1/2}_{t_{j-1}}\Delta t_i.$$

Since the function $\Lambda_t^{1/2}$ is integrable and g' is bounded, this implies that the sum of absolute values of the expectations of the cross terms decreases to zero as $n^{-1/2}$ and, hence, $n(R^n_{22})^2=O(n^{-1/2})$.

4. We verify that $nE(R^n_{23})^2\to 0$. Recall that

$$E(S_{t_i}-S_{t_{i-1}})^{2m}\le c_m(\Delta t_i)^m.$$

Using (31) we obtain the bound

$$E\widehat{C}^2_{xt}(t_{i-1},S_{t_{i-1}})(\Delta t_i)^2(S_{t_i}-S_{t_{i-1}})^2\le c\frac{(\Delta t_i)^3}{(1-t_{i-1})^{3/2}}.$$

To estimate the terms coming from the residual term of the Taylor expansion we use the Cauchy–Schwarz inequality and the bounds (22)–(24). This yields in the following:

$$E\widehat{C}^2_{xxx}(\tilde{t}_{i-1},\tilde{S}_{t_{i-1}})(S_{t_i}-S_{t_{i-1}})^6\le c\frac{(\Delta t_i)^3}{(1-t_i)^2},$$

$$E\widehat{C}^2_{xxt}(\tilde{t}_{i-1},\tilde{S}_{t_{i-1}})(S_{t_i}-S_{t_{i-1}})^4(\Delta t_i)^2\le c\frac{(\Delta t_i)^4}{(1-t_i)^3},$$

$$E\widehat{C}^2_{xtt}(\tilde{t}_{i-1},\tilde{S}_{t_{i-1}})(\Delta t_i)^4(S_{t_i}-S_{t_{i-1}})^2\le c\frac{(\Delta t_i)^5}{(1-t_i)^4}.$$

We dominate the L^2-norm of $n^{1/2}R^n_{23}$ by the sum of the L^2-norm of the random variables $[...]_i(S_{t_i} - S_{t_{i-1}})$, where $[...]_i$ is defined in (17). Taking into account that $\widehat{C}_{xx}(t,x) > 0$ and using the inequality $||a| - |b|| \le |a - b|$ we can write that

$$\|[...]_i(S_{t_i} - S_{t_{i-1}})\|_{L^2} \le c\Big(\|\widehat{C}_{xt}(t_{i-1}, S_{t_{i-1}})(t_i - t_{i-1})(S_{t_i} - S_{t_{i-1}})\|_{L^2} + ...\Big)$$

where we denote by dots the L^2-norms of the residual term in the first order Taylor expansion of the difference $\widehat{C}_x(t_i, S_{t_i}) - \widehat{C}_x(t_{i-1}, S_{t_{i-1}})$. Summing up and using the above estimates we conclude, applying Lemma 2.3, that the right-hand side of the above inequality tends to zero as $n \to \infty$ and we conclude.

5. It remains to check that $nE(R^n_{24})^2 \to 0$ and this happens to be the most delicate part of the proof. The expression for $nE(R^n_{24})^2$ involves the sum of expectations of squared terms and the sum of expectations of cross terms which we analyze separately.

For the squared terms the arguments are relatively straightforward. We apply the Ito formula to the function $\widehat{C}_x(t,x)$. Using the positivity of $\widehat{C}_{xx}(t,x)$ and the inequality $||a| - |b|| \le |a - b|$ we dominate the absolute value of the square bracket $[...]_i$ in the definition of R^n_{24}, given by the formula (17), by the absolute value of

$$\int_{t_{i-1}}^{t_i} (\widehat{C}_{xx}(t_{i-1}, S_{t_{i-1}}) - \widehat{C}_{xx}(t, S_t))dS_t - \int_{t_{i-1}}^{t_i} \Big(\widehat{C}_{xt}(t, S_t) + \frac{\sigma^2}{2} S_t^2 \widehat{C}_{xxx}(t, S_t)\Big)dt.$$

We check that

$$\sum_{i=1}^{n-1} ES^2_{t_{i-1}} \int_{t_{i-1}}^{t_i} (\widehat{C}_{xx}(t_{i-1}, S_{t_{i-1}}) - \widehat{C}_{xx}(t, S_t))^2 S_t^2 dt = O(n^{-1/4}), \tag{38}$$

$$\sum_{i=1}^{n-1} \Delta t_i ES^2_{t_{i-1}} \int_{t_{i-1}}^{t_i} (\widehat{C}^2_{xt}(t, S_t) + S_t^4 \widehat{C}^2_{xxx}(t, S_t))dt = O(n^{-1/2}). \tag{39}$$

A generic term of the first sum is dominated by

$$\Delta t_i E \sup_{t \le 1} S_t^4 \sup_{t_{i-1} \le t \le t_i} (\widehat{C}_{xx}(t, S_t) - \widehat{C}_{xx}(t_{i-1}, S_{t_{i-1}}))^2.$$

The Cauchy–Schwarz inequality allows us to separate the terms under the sign of expectation and reduce the problem to the estimation of the forth power of the difference $\widehat{C}_{xx}(t, S_t) - \widehat{C}_{xx}(t_{i-1}, S_{t_{i-1}})$. The Ito formula transforms this difference into the sum of a stochastic integral and an ordinary integral. Using consecutively the Burkholder and Cauchy–Schwarz inequalities and the bound (32) we have:

$$E \sup_{t\in[t_{i-1},t_i]} \left[\int_{t_{i-1}}^{t} \widehat{C}_{xxx}(u,S_u)S_u dS_u\right]^4 \le c_4 E\left[\int_{t_{i-1}}^{t_i} \widehat{C}^2_{xxx}(u,S_u)S_u^4 du\right]^2$$
$$\le c_4\Delta t_i E\int_{t_{i-1}}^{t_i} \widehat{C}^4_{xxx}(u,S_u)S_u^8 du$$
$$\le c\frac{(\Delta t_i)^2}{(1-t_i)^{7/2}}.$$

To estimate the ordinary integral we use the Jensen inequality for $f(x) = x^4$ and the bounds (33) and (34) and get that

$$E \sup_{t\in[t_{i-1},t_i]} \left[\int_{t_{i-1}}^{t} \Big(\widehat{C}_{xxt}(u,S_u) + \frac{1}{2}\sigma^2 S_u^2 \widehat{C}_{xxxx}(u,S_u)\Big)du\right]^4 \le c\frac{(\Delta t_i)^4}{(1-t_i)^{11/2}}.$$

Using these estimates we obtain that the sum in (38) is dominated, up to a multiplicative constant, by

$$\sum_{i=1}^{n-1}\left[\frac{(\Delta t_i)^2}{(1-t_i)^{7/4}} + \frac{(\Delta t_i)^3}{(1-t_i)^{11/4}}\right]$$

and the claimed asymptotics follows from Lemma 2.3.

Finally, similar arguments using the inequalities (31) and (32) give us the second asymptotic formula.

¿From the same estimates we obtain that

$$\sum_{i=1}^{n-1}(ES^2_{t_{i-1}}[...]_i^2)^{1/2} \le c\sum_{i=1}^{n-1}\frac{\Delta t_i}{(1-t_i)^{7/8}} + c\sum_{i=1}^{n-1}\frac{(\Delta t_i)^{3/2}}{(1-t_i)^{11/8}}.$$

The second sum in the right-hand side converges to zero while for the first one we can say only that it is dominated by a convergent integral. Using this observation we conclude that the sum of expectations of cross terms over indices i, j with $i < j$ and $t_j > a$ also can be done arbitrary small by choosing a sufficiently close to one.

Unexpectedly, the most difficult part of the proof is in establishing the convergence to zero of the sum of cross terms corresponding to the dates of revisions before $a < 1$, i.e. bounded away from the singularity.

Using the Taylor expansion we can reduce the problem to the case where the difference $\widehat{C}_x(t_i, S_{t_i}) - \widehat{C}_x(t_{i-1}, S_{t_{i-1}})$ is replaced by terms involving the derivatives $C_{xx}(t_{i-1}, S_{t_{i-1}})$, $C_{xt}(t_{i-1}, S_{t_{i-1}})$ and $C_{xxx}(t_{i-1}, S_{t_{i-1}})$.

To formulate the claim we introduce "reasonable" notations. Put

$$\alpha_i := \widehat{C}_{xx}(S_{t_{i-1}}, t_{i-1})S^2_{t_{i-1}}\Big(\frac{S_{t_i}}{S_{t_{i-1}}} - 1\Big),$$

$$\beta_i := S_{t_{i-1}}\widehat{C}_{xt}(S_{t_{i-1}}, t_{i-1})\Delta t_i + \frac{1}{2}S^3_{t_{i-1}}\widehat{C}_{xxx}(S_{t_{i-1}}, t_{i-1})\Big(\frac{S_{t_i}}{S_{t_{i-1}}} - 1\Big)^2,$$

$\gamma_i := |\alpha_i + \beta_i| - |\alpha_i|$. Let us define also the random variable $\chi_i := \operatorname{sign}(\alpha_i \beta_i)$ and the set $A_i := \{|\beta_i| < |\alpha_i|\}$.

The assertion needed to conclude is the lemma below. It is based on asymptotic analysis of expectations of some Gaussian integrals which are given in the next section and the following identities:

$$\begin{aligned} |\alpha + \beta| - |\alpha| &= |\beta|\chi I_A + |\beta| I_{\{\chi>0\}} I_{A^c} + (|\beta| - 2|\alpha|) I_{\{\chi\le 0\}} I_{A^c} \\ &= |\beta|\chi + 2(|\beta| - |\alpha|) I_{\{\chi\le 0\}} I_{A^c} - |\beta| I_{\{\chi=0\}} I_{A^c} \end{aligned}$$

where α, β are arbitrary random variables, $\chi := \operatorname{sign}(\alpha\beta)$, $A := \{|\beta| < |\alpha|\}$.

Lemma 2.5. *For every fixed $a \in]0,1[$*

$$\Big| \sum_{i<j,\, t_j \le a} E\gamma_i \gamma_j \Big| = o(1), \qquad n \to \infty.$$

Proof. The routine estimation $|E\gamma_i\gamma_j| \le E|\gamma_i||\gamma_j|$ does not work in our case. But for $i < j$

$$|E\gamma_i\gamma_j| = |E(\gamma_i E(\gamma_j|\mathcal{F}_{t_{j-1}}))| \le E(|\gamma_i||E(\gamma_j|\mathcal{F}_{t_{j-1}})|) \le E(|\beta_i||E(\gamma_j|\mathcal{F}_{t_{j-1}})|).$$

According to the above identity,

$$|E(\gamma_j|\mathcal{F}_{t_{j-1}})| \le |E(|\beta_j|\chi_j|\mathcal{F}_{t_{j-1}})| + 2E(|\beta_j| I_{A_j^c}|\mathcal{F}_{t_{j-1}}).$$

Using Lemma 3.2 of the next section with $\eta_u = S_{t_j}/S_{t_{j-1}} - 1$, $u = (\Delta t_j)^{1/2}$, we dominate the first term in the right-hand side by

$$\kappa(S_{t_{i-1}}|\widehat{C}_{xt}(S_{t_{i-1}}, t_{i-1})| + S^3_{t_{i-1}}|\widehat{C}_{xxx}(S_{t_{i-1}}, t_{i-1})|)(\Delta t_j)^{3/2}$$

It is easily seen from the explicit formulae that the coefficients above when $t_j \le a$ can be dominated uniformly by $c_a(1 + \sup_{t\le 1} S_t)$, i.e. by a random variable having all moments. In the same range of indices we have the bound $E(\beta_i^2|\mathcal{F}_{t_{i-1}}) \le \zeta_a(\Delta t_i)^2$ where ζ_a a random variable having all moments. It follows from here that

$$\sum_{i<j,\, t_j \le a} E(|\beta_i||E(|\beta_j|\chi_j|\mathcal{F}_{t_{j-1}})|) = O(n^{-1/2}).$$

We estimate $P(A_j^c|\mathcal{F}_{t_{j-1}})$ applying Lemma 3.3 of the next section with

$$c_1(t_{j-1}) := \frac{S^3_{t_{j-1}}\widehat{C}_{xxx}(S_{t_{j-1}}, t_{j-1})}{S^2_{t_{j-1}}\widehat{C}_{xx}(S_{t_{j-1}}, t_{j-1})}, \qquad c_2(t_{j-1}) := \frac{S_{t_{j-1}}\widehat{C}_{xt}(S_{t_{j-1}}, t_{j-1})}{S^2_{t_{j-1}}\widehat{C}_{xx}(S_{t_{j-1}}, t_{j-1})},$$

and $c(t_{j-1}) := 2(|c_1(t_{j-1})| + |c_2(t_{j-1})| + 1)$. On the interval $[0, a]$ the continuous process $c(t)$ can be dominated by a random variable ξ_a. Fix $\varepsilon > 0$ and choose N such that $P(\xi_a > N) < \varepsilon$. Lemma 3.3 implies that

$$P(A_j^c|\mathcal{F}_{t_{j-1}}) \le L_N(\Delta t_j)^{1/2} I_{\{c(t_{j-1})\}\le N} + I_{\{c(t_{j-1})>N\}}$$

and, therefore, $P(A_j^c) \le L_N(\Delta t_j)^{1/2} + \varepsilon \le 2\varepsilon$ when n is large enough. Using the Cauchy–Schwarz and Jensen inequalities we get that

$$\begin{aligned}\sum_{i<j,\, t_j\le a} E(|\beta_i||E(|\beta_j|I_{A_j^c}|\mathcal{F}_{t_{j-1}})|) &\le \sum_{t_i\le a}(E\beta_i^2)^{1/2}\sum_{t_j\le a}(E\beta_i^4)^{1/4}(P(A_j^c))^{1/4} \\ &\le (2\varepsilon)^{1/4}\sum_{t_i\le a}(E\beta_i^2)^{1/2}\sum_{t_j\le a}(E\beta_j^4)^{1/4}.\end{aligned}$$

Note that both sums in the right-hand side are bounded due to the inequalities $E\beta_j^2 \le \kappa(\Delta t_i)^2$ and $E\beta_j^4 \le \kappa(\Delta t_i)^4$. By the choice of ε the right-hand side can be made arbitrarily small. Thus, $nE(R_{24}^n)^2 \to 0$. □

3. Asymptotics of Gaussian Integrals

Let $\xi \in \mathcal{N}(0, 1)$ and let $\eta_u := e^{u\xi - \frac{1}{2}u^2} - 1$, $u \in [0, 1]$.

Lemma 3.1. *The following asymptotical formulae holds as $u \to 0$:*

$$\begin{aligned}E[\eta_u^2 - \eta_{-u}^2]I_{\{\eta_u>0\}} &= \frac{2}{\sqrt{2\pi}}u^3 + O(u^4),\\ E\eta_u^2\operatorname{sign}\eta_u &= \frac{2}{\sqrt{2\pi}}u^3 + O(u^4),\\ E\operatorname{sign}\eta_u &= -\frac{1}{\sqrt{2\pi}}u + O(u^3).\end{aligned}$$

Proof. Put

$$Z(u) := (e^{u\xi-\frac{1}{2}u^2} - 1)^2 - (e^{-u\xi-\frac{1}{2}u^2} - 1)^2.$$

Then $Z(0) = Z'(0) = Z''(0) = 0$, $Z'''(0) = 12(\xi^3 - \xi)$, and the function $Z^{(4)}(u)$ is bounded by a random variable having moments of any order. Using the Taylor formula we obtain that

$$EZ(u)I_{\{\xi\ge\frac{1}{2}u\}} = 2u^3E(\xi^3 - \xi)I_{\{\xi\ge\frac{1}{2}u\}} + O(u^4), \qquad u \to 0,$$

and we obtain the first formula. The second formula is a corollary of the first one since

$$E\eta_u^2\operatorname{sign}\eta_u = EZ(u)I_{\{\xi\ge\frac{1}{2}u\}} - E\eta_u^2 I_{\{|\xi|\le\frac{1}{2}u\}}$$

and the last term is $O(u^4)$ as $u \to 0$. Finally,

$$
\begin{aligned}
E\operatorname{sign}\eta_u = P(\xi > u/2) - P(\xi < u/2) &= 2(\Phi(0) - \Phi(u/2)) \\
&= -\frac{1}{\sqrt{2\pi}}u + \frac{1}{4}\varphi(\tilde{u})\tilde{u}u^2,
\end{aligned}
$$

where $\tilde{u} \in [0, u/2]$. □

Lemma 3.2. *There exists a constant κ such that for any real A*

$$
\left|E|\eta_u^2 - Au^2|\operatorname{sign}(\eta_u^2 - Au^2)\eta_u\right| \leq \kappa(1 + |A|)u^3. \tag{40}
$$

Proof. Note that $|x| \operatorname{sign} xy = x \operatorname{sign} y$. Therefore the left-hand side of (40) is dominated by

$$
\left|E\eta_u^2 \operatorname{sign}\eta_u\right| + |A|u^2\left|E\operatorname{sign}\eta_u\right|
$$

and the result holds by virtue of the previous lemma. □

Lemma 3.3. *For every $N > 0$ there is a constant L_N such that for all $u \in [0, 1]$*

$$
P(|c_1\eta_u^2 + c_2u^2| > |\eta_u|) \leq L_N I_{\{c \leq N\}} u + I_{\{c > N\}}.
$$

for any constants c_1, c_2 and $c := 2(|c_1| + |c_2| + 1)$.

Proof. Suppose that $N \geq c > 2$, the only case where the work is needed. It is easy to see that

$$
\begin{aligned}
P(|c_1\eta_u^2 + c_2u^2| > |\eta_u|) &\leq P((c/2)\eta_u^2 + (c/2)u^2 > |\eta_u|) \\
&\leq P(c|\eta_u| > 1) + P(|\eta_u| < cu^2).
\end{aligned}
$$

The probabilities in the right-hand side as functions of c are increasing and it remains to dominate their values at the point $c = N$. The required bound holds for the first probability in the right-hand side (and even with a constant which does not depend on N). Indeed, using the Chebyshev inequality, finite increments formula, and the bound $\varphi(x) \leq 1/\sqrt{2\pi}$ we have:

$$
P(N|\eta_u| > 1) \leq \frac{1}{N}E|\eta_u| \leq \frac{1}{2}E|\eta_u| = \Phi(u/2) - \Phi(-u/2) \leq \frac{1}{\sqrt{2\pi}}u.
$$

For $u \geq 1/\sqrt{2N}$ the second probability is dominated by linear functions with $L_N \geq \sqrt{2N}$. For $u < 1/\sqrt{2N}$ we write it as

$$
P(u/2 \leq \xi < (1/u)\ln(1 + Nu^2) + u/2) + P((1/u)\ln(1 - Nu^2) + u/2 < \xi < u/2).
$$

Using again the finite increments formula we obtain that

$$
P(u/2 \leq \xi < (1/u)\ln(1 + Nu^2) + u/2) \leq \frac{1}{\sqrt{2\pi}}Nu.
$$

On the interval $]0, 1/\sqrt{2N}[$ we have the bound $(1/u)\ln(1 - Nu^2) \geq -\kappa Nu$ where $\kappa > 0$ is the maximum of the function $-\ln(1 - x)/x$ on the interval $]0, 1/2]$. It follows that

$$P((1/u)\ln(1 - Nu^2) + u/2 < \xi < u/2) \leq \frac{1}{\sqrt{2\pi}}\kappa Nu.$$

Thus, the second probability also admits a linear majorant on the whole interval $[0, 1]$. □

Acknowledgement

The authors express their thanks to Emmanuel Denis and the anonymous referee for constructive criticism and helpful suggestions.

References

1. Geiss, S., Quantitative approximation of certain stochastic integrals, *Stochastics and Stochastic Reports*, **73**, (2002), 3–4, 241–270.
2. Granditz, P. and Schachinger, W., Leland's approach to option pricing: The evolution of discontinuity, *Mathematical Finance*, **11**, (2001), 3, 347–355.
3. Grannan, E. R. and Swindle, G. H., Minimizing transaction costs of hedging strategies, *Mathematical Finance*, **6**, (1996), 4, 341–364.
4. Denis, E. and Kabanov, Yu. M., *Functional limit theorem for Leland–Lott hedging strategy*, Working paper.
5. Kabanov, Yu. M. and Safarian, M., On Leland's strategy of option pricing with transaction costs, *Finance and Stochastics*, **1**, 3, (1997), 239–250.
6. Leland, H., Option pricing and replication with transactions costs, *Journal of Finance*, **XL**, (1985), 5, 1283–1301.
7. Lott, K., *Ein Verfahren zur Replikation von Optionen unter Transaktionkosten in stetiger Zeit*, (1993), Dissertation, Universität der Bundeswehr München. Institut für Mathematik und Datenverarbeitung.
8. Pergamenshchikov, S., Limit theorem for Leland's strategy, *The Annals of Applied Probability*, **13**, (2003), 1099–1118.
9. Sekine, J. and Yano, J., *Hedging errors of Leland's strategies with time-inhomogeneous rebalancing*, Preprint.
10. Zariphopoulou, T., Stochastic control methods in asset pricing, in *Handbook of Stochastic Analysis and Applications*, edited by Kannan, D. and Lakshmikantham, (2002), V. Marcel Dekker, New York–Basel.

Variance Reduction for MC/QMC Methods to Evaluate Option Prices

Jean-Pierre Fouque*, Chuan-Hsiang Han†, and Yongzeng Lai‡

[1]Department of Statistics and Applied Probability, University of California, Santa Barbara
E-mail: fouque@pstat.ucsb.edu.*
[2]Department of Quantitative Finance, National Tsing Hua University
E-mail: chhan@mx.nthu.edu.tw.†
[3]Department of Mathematics, Wilfrid Laurier University
E-mail: ylai@wlu.ca.‡

Several variance reduction techniques including importance sampling, (martingale) control variate, (randomized) Quasi Monte Carlo method, QMC in short, and some possible combinations are considered to evaluate option prices. By means of perturbation methods to derive some option price approximations, we find from numerical results in Monte Carlo simulations that the control variate method is more efficient than importance sampling to solve European option pricing problems under multifactor stochastic volatility models. As an alternative, QMC method also provides better convergence than basic Monte Carlo method. But we find an example where QMC method may produce erroneous solutions when estimating the low-biased solution of an American option. This drawback can be effectively fixed by adding a martingale control to the estimator adopting Quasi random sequences so that low-biased estimates obtained are more accurate than results from Monte Carlo method. Therefore by taking advantages of martingale control variate and randomized QMC, we find significant improvement on variance reduction for pricing derivatives and their sensitivities. This effect should be understood as that martingale control variate plays the role of a smoother under QMC method to permit better convergence.
Key words: Importance Sampling, Martingale Control Variate, Randomized Quasi Monte Carlo, Perturbation approximation.

1. Introduction

Monte Carlo method and Quasi Monte Carlo method (MC/QMC method in short) are important tools for integral problems in computational finance. They are popularly applied particularly in cases of solutions without closed form, for example American put option prices under the Black-Scholes model and European option prices under multi-factor stochastic volatility models, both which we will consider in the present paper. Our goal is to find an efficient variance reduction method to improve the convergence of MC/QMC methods.

The study of variance reduction methods for option pricing problems has been very fruitful during the last two decades [7]. They are important, just to name a few, for computing *greeks* (sensitivities of options prices with respect to model variables or parameters), risk management and model calibration. Due to the complexity of financial derivatives and pricing models, it is difficult to find one general approach to reduce variances of associated MC/QMC methods. However in [2] and [3], we can use (local) martingale control variate methods to evaluate European, Barrier and American options through Monte Carlo simulations. This paper concerns numerical comparisons with several variance reduction techniques such as control variate, importance sampling, Brownian bridge, randomized QMC and their possible combinations.

Our goal is to explore an efficient MC/QMC method to evaluate American put options under Black-Scholes model and European call options under multi-factor stochastic volatility models. Firstly, we are motivated by previous results in [1] and [2] where importance sampling method and martingale control variate methods are used respectively under Monte Carlo simulations for European option pricing problems. From many numerical comparisons between these two methods, we find that martingale control variate method performs much better than importance sampling in terms of variance reduction power. Secondly, we investigate the efficiency of QMC method for option pricing. As an integration method using quasi-random sequences (also called low-discrepancy sequences), QMC method [7] has better convergence rates than Monte Carlo method, see Section 4, under appropriate dimensionality and regularity of the integrand. However in many financial applications, these conditions are not satisfied. We give a counterexample in Section 4.2 that shows that using QMC method such as Niderreiter or Sobol sequences, gives erroneous estimates for low-biased solutions of American put options. To our best knowledge, this is the first counterexample showing the failure of applying QMC method in financial applications. However, when we combine a martingale control variate with QMC method, very accurate low-biased

*Work partially supported by NSF grant DMS-0455982.

†Work supported by NSC grant 95-2115-M-007-017-MY2, Taiwan, and National Center for Theoretical Sciences (NCTS), Taiwan.

‡Work partially supported by an Natural Sciences and Engineering Research Council (NSERC) of Canada grant.

estimates are obtained compared to Monte Carlo method, see Table 7 for details. In brief, when the regularity of the estimator is not good enough, using QMC method can be problematic. The effect of martingale control variate thus plays the role of a smoother which improves the regularity of the controlled estimator.

Because of the mentioned benefits of martingale control variate and its combination with QMC method, we continue to explore in detail the effect of martingale control variate with randomized QMC method for the European option pricing problems. Under multifactor stochastic volatility models, the dimension of the randomized QMC method becomes high. Typically in our experiments the dimension goes up to 300. This can be an obstacle for QMC method to reduce variance in a significant way as its convergence rate depends on the dimension. Based on our experiments in Section 5, the effect of martingale control variate again tremendously improves the regularity of the controlled estimator. Compared to numerical results from the basic Monte Carlo method, randomized QMC method improves the variance only by a single digit, while martingale control variate under Monte Carlo simulations improves variance by 50 times. The combination of the martingale control variate with the randomized QMC method improves variance reduction ratios up to 700 times.

The organization of this paper is as follows. In Section 2, we introduce stochastic volatility models and European option price approximations obtained from [5] by means of singular and regular perturbation methods. Section 3 reviews two variance reduction methods, namely control variate and importance sampling, and compare their variance reduction performances. In Section 4, we introduce the QMC method and show a counterexample where the method fails. We then show how to combine this method with a correction by a martingale control variate. Section 5 tests several combinations of martingale control variate methods with and without randomized QMC method, including the Sobol sequence and L'Ecuyer type good lattice points together with the Brownian bridge sampling technique. We also consider option prices and their *deltas*, first-order partial derivative with respect to the underlying price.

2. Multi-Factor Stochastic Volatility Models and Option Price Approximations

Under the physical probability measure, a family of multi-factor stochastic volatility models evolves as

$$\begin{aligned} dS_t &= \kappa S_t dt + \sigma_t S_t dW_t^{(0)}, \\ \sigma_t &= f(Y_t, Z_t), \\ dY_t &= \alpha c_1(Y_t)dt + \sqrt{\alpha}\, g_1(Y_t)dW_t^{(1)} \\ dZ_t &= \delta c_2(Z_t)dt + \sqrt{\delta}\, g_2(Z_t)dW_t^{(2)}, \end{aligned}$$

where S_t is the underlying asset price with a constant rate of return κ and the random volatility σ_t, Y_t and Z_t are driving volatility processes varying with time scales $1/\alpha$ and $1/\delta$ respectively. The standard Brownian motions $\left(W_t^0, W_t^1, W_t^2\right)$ are possibly correlated as described below. The volatility function f is assumed bounded and bounded away from 0, and continuous with respect to its second variable z. The coefficient functions of Y_t, namely c_1 and g_1 are assumed to be chosen such that Y_t is an ergodic diffusion. The Ornstein-Uhlenbeck (OU) process is a typical example by defining α the rate of mean-reversion, $c_1(y) = m_1 - y$, and $g_1(y) = \nu_1\sqrt{2}$, where m_1 is the long-run mean and ν_1 is the long-run standard deviation, such that $\Phi = \mathcal{N}(m, \nu^2)$ is the invariant distribution. The coefficient functions of Z_t, namely c_2 and g_2 are assumed to satisfy the existence and uniqueness conditions of diffusions [18]. For simplicity, we set the process Z_t to be another OU process by choosing $c_2(z) = m_2 - z$, and $g_2(z) = \nu_2\sqrt{2}$, where m_2 is the long-run mean and ν_2 is the long-run standard deviation. Suppose $\left(W_t^0, W_t^1, W_t^2\right)$ are correlated according to the following cross-variations:

$$
\begin{aligned}
d < W^{(0)}, W^{(1)} >_t &= \rho_1 dt, \\
d < W^{(0)}, W^{(2)} >_t &= \rho_2 dt, \\
d < W^{(1)}, W^{(2)} >_t &= \left(\rho_1\rho_2 + \sqrt{1-\rho_1^2}\rho_{12}\right) dt,
\end{aligned}
$$

where the instant correlations ρ_1,ρ_2, and ρ_{12} satisfy $|\ \rho_1\ | < 1$ and $|\ \rho_2^2 + \rho_{12}^2\ | < 1$ respectively.

Under the risk-neutral probability measure $\boldsymbol{P}^{\star}$, a family of multi-factor SV models can be described as follows

$$
\begin{aligned}
dS_t &= rS_t dt + \sigma_t S_t dW_t^{(0)\star}, \\
\sigma_t &= f(Y_t, Z_t), \\
dY_t &= \left(\alpha(m_1 - Y_t) - \nu_1\sqrt{2\alpha}\Lambda_1(Y_t, Z_t)\right) dt \\
&\quad + \nu_1\sqrt{2\alpha}\left(\rho_1 dW_t^{(0)\star} + \sqrt{1-\rho_1^2} dW_t^{(1)\star}\right), \\
dZ_t &= \left(\delta(m_2 - Z_t) - \nu_2\sqrt{2\delta}\Lambda_2(Y_t, Z_t)\right) dt \\
&\quad + \nu_2\sqrt{2\delta}\left(\rho_2 dW_t^{(0)\star} + \rho_{12} dW_t^{(1)\star} + \sqrt{1-\rho_2^2-\rho_{12}^2} dW_t^{(2)\star}\right),
\end{aligned}
\tag{2.1}
$$

where $\left(W_t^{(0)\star}, W_t^{(1)\star}, W_t^{(2)\star}\right)$ are independent standard Brownian motions. The risk-free interest rate of return is denoted by r. The functions Λ_1 and Λ_2 are the combined market prices of risk and volatility risk, they are assumed to be bounded and dependent only on the variables y and z. The process (S_t, Y_t, Z_t) is Markovian.

The payoff of an European-style option is an integrable function, say H, of the stock price S_T at the maturity date T. The price of this option is defined as the expectation of the discounted payoff conditioned on the current stock price and driving volatility levels due to the Markov property of the joint dynamics (2.1). By introducing the notation $\varepsilon = 1/\alpha$, the European option price is given by

$$P^{\varepsilon,\delta}(t,x,y,z) = \boldsymbol{E}^{\star}\left\{e^{-r(T-t)}H(S_T) \mid S_t = x, Y_t = y, Z_t = z\right\}. \tag{2.2}$$

2.1 Vanilla European Option Price Approximations

By an application of Feynman-Kac formula, $P^{\varepsilon,\delta}(t,x,y,z)$ defined in (2.2) can also be represented by solving the three-dimensional partial differential equation

$$\begin{aligned} &\frac{\partial P^{\varepsilon,\delta}}{\partial t} + \mathcal{L}^{\varepsilon,\delta}_{(S,Y,Z)} P^{\varepsilon,\delta} - r P^{\varepsilon,\delta} = 0, \\ &P^{\varepsilon,\delta}(T,x,y,z) = H(x), \end{aligned} \tag{2.3}$$

where $\mathcal{L}^{\varepsilon,\delta}_{(S,Y,Z)}$ denotes the infinitesimal generator of the Markovian process (S_t, Y_t, Z_t) given by (2.1). Assuming that the parameters ε and δ are small, $0 < \varepsilon, \delta \ll 1$, Fouque *et al.* in [5] use a combination of regular and singular perturbation methods to derive the following pointwise option price approximation

$$P^{\varepsilon,\delta}(t,x,y,z) \approx \tilde{P}(t,x,z),$$

where

$$\begin{aligned} \tilde{P} = {} & P_{BS} \\ & + (T-t)\left(V_0\frac{\partial}{\partial\sigma} + V_1 x\frac{\partial^2}{\partial x\partial\sigma} + V_2 x^2\frac{\partial^2}{\partial x^2} + V_3 x\frac{\partial}{\partial x}\left(x^2\frac{\partial^2}{\partial x^2}\right)\right)P_{BS}, \end{aligned} \tag{2.4}$$

with an accuracy of order $(\varepsilon|\log\varepsilon| + \delta)$ for call options. The leading order price $P_{BS}(t,x;\bar{\sigma}(z))$ is independent of the y variable and is the homogenized price which solves the Black-Scholes equation

$$\begin{aligned} \mathcal{L}_{BS}(\overline{\sigma}(z))P_{BS} &= 0, \\ P_{BS}(T,x;\bar{\sigma}(z)) &= H(x). \end{aligned}$$

Here the z-dependent **effective volatility** $\overline{\sigma}(z)$ is defined by

$$\overline{\sigma}^2(z) = \langle f^2(\cdot,z)\rangle, \tag{2.5}$$

where the brackets denote the average with respect to the invariant distribution $\mathcal{N}(m_1, \nu_1^2)$ of the fast factor (Y_t). The parameters (V_0, V_1, V_2, V_3) are given by

$$V_0 = -\frac{\nu_2\sqrt{\delta}}{\sqrt{2}}\langle\Lambda_2\rangle\overline{\sigma}', \tag{2.6}$$

$$V_1 = \frac{\rho_2\nu_2\sqrt{\delta}}{\sqrt{2}}\langle f\rangle\overline{\sigma}', \tag{2.7}$$

$$V_2 = \frac{\nu_1\sqrt{\varepsilon}}{\sqrt{2}}\left\langle\Lambda_1\frac{\partial\phi}{\partial y}\right\rangle, \tag{2.8}$$

$$V_3 = -\frac{\rho_1\nu_1\sqrt{\varepsilon}}{\sqrt{2}}\left\langle f\frac{\partial\phi}{\partial y}\right\rangle, \tag{2.9}$$

where $\overline{\sigma}'$ denotes the derivative of $\bar{\sigma}$, and the function $\phi(y,z)$ is a solution of the Poisson equation

$$\mathcal{L}_0\phi(y,z) = f^2(y,z) - \overline{\sigma}^2(z).$$

The parameters V_0 and V_1 (resp. V_2 and V_3) are small of order $\sqrt{\varepsilon}$ (resp. $\sqrt{\delta}$). The parameters V_0 and V_2 reflects the effect of the market prices of volatility risk. The parameters V_1 and V_3 are proportional to the correlation coefficients ρ_2 and ρ_1 respectively. In [5], these parameters are calibrated using the observed implied volatilities. In the present work, the model (2.1) will be fully specified, and these parameters are computed using the formulas above.

3. Monte Carlo Simulations: Two Variance Reduction Methods

In this section, two variance reduction methods, namely importance sampling [1] and control variates [2], to evaluate European option prices by Monte Carlo simulations are compared under multi-factor stochastic volatility models. The technique of importance sampling has been introduced to evaluate European and Asian option prices in [1, 6]. We briefly review this methodology in Section 3.1. A control variate method based on [2] is reviewed in Section 3.2. This method has been applied to several option pricing problems including Barrier and American options [3]. In Section 3.3, test examples of one and two factor stochastic volatility models are demonstrated to show that the control variate method performs better than importance sampling in terms of variance reduction power.

To simplify notations, we present the stochastic volatility model in (2.1) in the vector form

$$dV_t = b(t, V_t)dt + a(t, V_t)d\eta_t, \tag{3.10}$$

where we set

$$v = \begin{pmatrix} x \\ y \\ z \end{pmatrix}, \quad V_t = \begin{pmatrix} S_t \\ Y_t \\ Z_t \end{pmatrix}, \quad \eta_t = \begin{pmatrix} W_t^{(0)\star} \\ W_t^{(1)\star} \\ W_t^{(2)\star} \end{pmatrix},$$

and we define the drift

$$b(t,v) = \begin{pmatrix} rx \\ \alpha(m_1 - y) - \nu_1 \sqrt{2\alpha}\, \Lambda_1(y,z) \\ \delta(m_2 - z) - \nu_2 \sqrt{2\delta}\, \Lambda_2(y,z) \end{pmatrix},$$

and the diffusion matrix

$$a(t,v) = \begin{pmatrix} f(y,z)x & 0 & 0 \\ \nu_1 \sqrt{2\alpha}\rho_1 & \nu_1 \sqrt{2\alpha}\sqrt{1-\rho_1^2} & 0 \\ \nu_2 \sqrt{2\delta}\rho_2 & \nu_2 \sqrt{2\delta}\rho_{12} & \nu_2 \sqrt{2\delta}\sqrt{1-\rho_2^2-\rho_{12}^2} \end{pmatrix}.$$

The price $P(t,x,y,z)$ of an European option at time t is given by

$$P(t,v) = \boldsymbol{E}^\star \left\{ e^{-r(T-t)} H(S_T) \mid V_t = v \right\}. \tag{3.11}$$

The basic Monte Carlo simulation estimates the option price $P(0, S_0, , Y_0, Z_0)$ at time 0 by the sample mean

$$\frac{1}{N} \sum_{k=1}^{N} e^{-rT} H(S_T^{(k)}) \tag{3.12}$$

where N is the total number of sample paths and $S_T^{(k)}$ denotes the k-th simulated stock price at time T.

3.1 Importance Sampling

A change of drift in the model dynamics (3.10) can be obtained by

$$dV_t = (b(t,V_t) - a(t,V_t)h(t,V_t))\,dt + a(t,V_t)d\tilde{\eta}_t, \tag{3.13}$$

where

$$\tilde{\eta}_t = \eta_t + \int_0^t h(s,V_s)ds.$$

The instantaneous shift $h(s,V_s)$ is assumed to satisfy the Novikov's condition

$$\boldsymbol{E}^\star \left\{ \exp\left(\frac{1}{2} \int_0^T h^2(s,V_s)ds \right) \right\} < \infty.$$

By Girsanov Theorem, one can construct the new probability measure $\tilde{\boldsymbol{P}}$ by

$$\frac{d\boldsymbol{P}^\star}{d\tilde{\boldsymbol{P}}} = Q_T,$$

where the Radon-Nikodyn derivative is defined as

$$Q_T = \exp\left(\left\{ \int_0^T h(s,V_s)d\tilde{\eta}_s - \frac{1}{2} \int_0^T \|h(s,V_s)\|^2 ds \right\} \right), \tag{3.14}$$

such that $\tilde{\eta}_t$ is a Brownian motion under $\tilde{\boldsymbol{P}}$. The option price P can be written as

$$P(t,v) = \tilde{\boldsymbol{E}}\left\{e^{-r(T-t)}H(S_T)Q_T \mid V_t = v\right\}. \tag{3.15}$$

By an application of Ito's formula to $P(t, V_t)\, Q_t$, one could obtain a zero variance of the discounted payoff $e^{-r(T-t)}H(S_T)Q_T$ by optimally choosing

$$h = -\frac{1}{P}\left(a^T \nabla P\right). \tag{3.16}$$

(See details in [1, 15].) The super script notation T denotes transpose and ∇ denotes the gradient. However, neither the price P nor its gradient ∇P were known in advance.

The idea of importance sampling techniques introduced in [1] is to approximate unknown option price $P^{\varepsilon,\delta}$ by $\tilde{P}$ as in (2.4). Then the Monte Carlo simulations are done under the new measure $\tilde{\boldsymbol{P}}$:

$$P(t,x,y,z) \approx \frac{1}{N}\sum_{k=1}^{N} e^{-r(T-t)}H(S_T^{(k)})Q_T^{(k)}, \tag{3.17}$$

where N is the total number of simulations, and $S_T^{(k)}$ and $Q_T^{(k)}$ denote the final value of the k-th realized trajectory (3.13) and weight (3.14) respectively.

3.2 Control Variate Method

A control variate with m multiple controls is defined as:

$$P^{CV} \triangleq P^{MC} + \sum_{i=1}^{m} \lambda_i(\hat{P}_C^i - P_C^i). \tag{3.18}$$

We denote by P^{MC} the sample mean of outputs from an IID simulation procedure. Each $\hat{P}_C^i$ represents the sample mean of those outputs jointly distributed by the previous simulation procedure. In addition, we assume $\hat{P}_C^i$ has the mean P_C^i which at best has a closed-form expression in order to reduce computational cost. The control variate P^{CV} thus becomes an unbiased estimator of P^{MC}. Each control parameter λ_i needs to be chosen to minimize the variance of P^{CV} as the coefficients in least squares regression. A detailed discussion on control variates can be found in [7] and an application to Asian option option in [8].

A constructive way to build control variate estimators under diffusion models (2.1) is as follows. Based on Ito's formula, the discounted option price satisfies

$$de^{-rs}P(s,S_s,Y_s,Z_s) = e^{-rs}\left(\frac{\partial}{\partial t} + \mathcal{L}_{(S,Y,Z)}^{\varepsilon,\delta} - r\cdot\right)P\,ds + e^{-rs}\,\nabla P\cdot(a\cdot d\eta_s).$$

The first term on the right hand side is crossed out because of (2.3). Integrating above equation in time between the current time t and the expiry date T, and using the terminal condition $P(T, S_T, Y_T, Z_T) = H(S_T)$,

$$(3.19) \qquad P(t, S_t, Y_t, Z_t) = e^{-r(T-t)} H(S_T) - \int_t^T e^{-r(s-t)} \nabla P \cdot (a \cdot d\eta_t)$$

is deduced.

However, the unknown price process $\{P(s, X_s, Y_s, Z_s)\}$ along trajectories between $\{t \le s \le T\}$ appears in each stochastic integral in (3.19). The use of price approximation (2.4)

$$P^{\varepsilon,\delta}(t, x, y, z) \approx \tilde{P}(t, x, z),$$

suggests a constructive way to build the control variate

$$(3.20) \quad P^{CV} = e^{-r(T-t)} H(S_T) - \int_t^T e^{-r(s-t)} \frac{\partial \tilde{P}}{\partial x} \sigma_s S_s dW_s^{(0)\star}$$
$$- \int_t^T e^{-r(s-t)} \frac{\partial \tilde{P}}{\partial z} \nu_2 \sqrt{2\delta} \left(\rho_2 dW_s^{(0)\star} + \rho_{12} dW_s^{(1)\star} + \sqrt{1 - \rho_2^2 - \rho_{12}^2} dW_s^{(2)\star} \right)$$

because $\tilde{P}$ is independent of the variable y. It is readily observed that P^{CV} is unbiased since by the martingale property of the stochastic integrals, the conditional expectation of the stochastic integrals are zero. In addition, it naturally suggests multiple estimators of optimal control parameters. To fit in the setup of the control variate with multiple controls (3.18), we have chosen for $i \in \{1, 2\}$:

$$P^{MC} = e^{-r(T-t)} H(S_T)$$
$$\lambda_i = -1$$
$$\hat{P}_C^i = \int_t^T e^{-r(s-t)} \frac{\partial \tilde{P}}{\partial x} \sigma_s S_s dW_s^{(0)\star} I_{\{i=1\}}$$
$$+ \int_t^T e^{-r(s-t)} \frac{\partial \tilde{P}}{\partial z} \nu_2 \sqrt{2\delta} \left(\rho_2 dW_s^{(0)\star} + \rho_{12} dW_s^{(1)\star} + \sqrt{1 - \rho_2^2 - \rho_{12}^2} dW_s^{(2)\star} \right) I_{\{i=2\}}$$
$$P_C^i = 0.$$

3.3 Numerical Results

Two sets of numerical experiments are proposed in order to compare the variance reduction performances of importance sampling and control variate described previously. The first set of experiments is for one-factor SV models and the second set is for two-factor SV models. These experiments are done only for vanilla European call options.

3.3.1 One-Factor SV Models

Under the framework of the two-factor SV model (2.1), an one-factor SV model is obtained by setting all parameters as well as the initial condition used to describe the second factor Z_t in (2.1) to zero. Our test model is chosen as in Fouque and Tullie [6], in which they used an Euler scheme to discretize the diffusion process V_t to run the Monte Carlo simulations. The time step is 10^{-3} and the number of realizations is 10000.

The one-factor stochastic volatility model is specified in Tables 1 and 2. In [6] the authors proposed an importance sampling technique by using an approximate option price obtained by a fast mean-reversion expansion. This approach is described in Section 3.1. Since only the one-factor SV model is considered, the zero-order price approximation reduces to

$$P_{BS}(t,x;\bar{\sigma}) = x\mathcal{N}(d_1(x)) - Ke^{-r(T-t)}\mathcal{N}(d_2(x)), \tag{3.21}$$

where

$$\begin{aligned} d_1(x) &= \frac{\ln(x/K) + (r + \frac{1}{2}\bar{\sigma}^2)(T-t)}{\bar{\sigma}\sqrt{T-t}}, \\ d_2(x) &= d_1(x) - \bar{\sigma}\sqrt{T-t}, \\ \mathcal{N}(d) &= \frac{1}{\sqrt{2\pi}}\int_{-\infty}^{d} e^{-u^2/2}du, \end{aligned}$$

the constant effective volatility $\bar{\sigma} = \bar{\sigma}(0)$ is defined in (2.5), and the first-order price approximation reduces to

$$\tilde{P} = P_{BS} + (T-t)\left(V_2 x^2 \frac{\partial^2}{\partial x^2} + V_3 x \frac{\partial}{\partial x}\left(x^2 \frac{\partial^2}{\partial x^2}\right)\right) P_{BS}.$$

In [6] it is found that the importance sampling technique performs best by employing the first-order price approximation $\tilde{P}$. According to different level of mean-reverting rate α, numerical results shown on Table 1 in [6] are copied to the second column of our Table 3, in which VAR_{MC} denotes the variance computed from basic Monte Carlo and $VAR^{IS}(\tilde{P})$ denotes the variance computed from importance sampling.

Our procedure to construct the control variate was described in Section 3.2. Since only one-factor model is considered, the control variate defined in (3.20) is reduced to

$$P^{CV} = e^{-r(T-t)}H(S_T) - \int_t^T e^{-r(s-t)} \frac{\partial P_{BS}}{\partial x} \sigma_s S_s dW_s^{(0)\star}. \tag{3.22}$$

Notice that we choose the zero-order option price approximation P_{BS} instead of the first-order price approximation $\tilde{P}$. The reason is that we have not found any major improvement by using $\tilde{P}$ instead of P_{BS} in our empirical results. In the third column of Table 3, we list the sample variance ratios obtained from

Table 1. Parameters used in the one-factor stochastic volatility model (2.1).

r	m_1	m_2	ν_1	ν_2	ρ_1	ρ_2	ρ_{12}	Λ_1	λ_2	$f(y,z)$
10%	-2.6	0	1	0	-0.3	0	0	0	0	$\exp(y)$

Table 2. Initial conditions and call option parameters.

$\$S_0$	Y_0	Z_0	$\$K$	T years
110	-2.32	0	100	1

Table 3. Comparison of estimated variance reduction ratio for European call options with various α's. Notation VAR_{MC} is the sample variance from basic Monte Carlo simulation. $VAR^{IS}(\tilde{P})$ is the sample variance computed from the important sampling with $\tilde{P}$ defined in (2.4) as an approximate option price. $VAR^{CV}(P_{BS})$ is the sample variance computed from the control variate with P_{BS} defined in (3.21) as an approximate option price.

α	$VAR_{MC}/VAR^{IS}(\tilde{P})$	$VAR_{MC}/VAR^{CV}(P_{BS})$
0.5	7.8095	49.8761
1	15.7692	29.9266
5	19.3333	31.8946
10	29.6250	51.4791
25	36.7143	100.1556
50	48.0000	183.6522
100	106.3333	229.9224

Table 4. Parameters used in the two-factor stochastic volatility model (2.1).

r	m_1	m_2	ν_1	ν_2	ρ_1	ρ_2	ρ_{12}	Λ_1	λ_2	$f(y,z)$
10%	-0.8	-0.8	0.5	0.8	-0.2	-0.2	0	0	0	$\exp(y+z)$

Table 5. Initial conditions and call option parameters.

$\$S_0$	Y_0	Z_0	$\$K$	T years
55	-1	-1	50	1

the basic Monte Carlo and the Monte Carlo with our control variate, namely $VAR_{MC}/VAR^{CV}(P_{BS})$. For each choice of α in Table 3, the computing times of the basic Monte Carlo method, importance sampling method, and control variate method take about 0.2, 2.8, 2.1 seconds, respectively when we implement them in Matlab 6 on a computer equipped with an Intel Centrino 2.2GHz CPU. From these test examples, the control variate given in (3.22) apparently dominates the importance sampling.

3.3.2 Two-Factor SV Models

We continue to investigate the performance of variance reduction for the two-factor SV model (2.1) defined in Table 4 and 5. Fouque and Han [1] present an

Table 6. Comparison of sample variances for various values of α and δ. Notation VAR_{MC} is the sample variance from basic Monte Carlo simulation, $VAR^{IS}(\tilde{P})$ is the sample variance computed from the important sampling with $\tilde{P}$ defined in (2.4) as an approximate option price. $VAR^{CV}(P_{BS})$ is the sample variance computed from the control variate with P_{BS} defined in (3.24) as an approximate option price.

α	δ	$VAR_{MC}/VAR^{IS}(\tilde{P})$	$VAR_{MC}/VAR^{CV}(P_{BS})$
5	1	13.4476	15.6226
20	0.1	17.5981	101.7913
50	0.05	32.9441	167.7985
100	0.01	24.4564	284.6705

importance sampling technique as described in Section 3.1 to evaluate European option prices. Their numerical results extracted from Table 3 in [1] are summarized as variance ratios in the third column of Table 6. According to different rates of mean-reversion α's and δ's for each factor, we illustrate ratios of sample variances computed from the basic Monte Carlo, denoted by VAR_{MC} and the Monte Carlo simulations with importance sampling, denoted by $VAR^{IS}(\tilde{P})$. Among these Monte Carlo simulations, there is a total of 5000 sample paths in (3.17) simulated based on the discretization of the diffusion process V_t using an Euler scheme with time step $\Delta t = 0.005$.

As in the case of one-factor SV models, we do not find apparent advantage of variance reduction by choosing the first-order approximate option price $\tilde{P}$ compared to using of the zero-order approximation P_{BS}. Hence the control variate implemented in this numerical experiment is given by

$$(3.23)\quad P^{CV} = e^{-r(T-t)}H(S_T) - \int_t^T e^{-r(s-t)}\frac{\partial P_{BS}}{\partial x}\sigma_s S_s dW_s^{(0)\star} - \int_t^T e^{-r(s-t)}\frac{\partial P_{BS}}{\partial z}\nu_2\sqrt{2\delta}\left(\rho_2 dW_s^{(0)\star}+\rho_{12}dW_s^{(1)\star}+\sqrt{1-\rho_2^2-\rho_{12}^2}\,dW_s^{(2)\star}\right),$$

where

$$(3.24)\quad \begin{aligned} P_{BS}(t,x;\bar{\sigma}(z)) &= x\mathcal{N}(d_1(x,z)) - Ke^{-r(T-t)}\mathcal{N}(d_2(x,z)),\\ d_1(x,z) &= \frac{\ln(x/K) + (r+\frac{1}{2}\bar{\sigma}^2(z))(T-t)}{\bar{\sigma}(z)\sqrt{T-t}},\\ d_2(x,z) &= d_1(x,z) - \bar{\sigma}(z)\sqrt{T-t}. \end{aligned}$$

In the fourth column of Table 6, we list the sample variance ratios obtained from the basic Monte Carlo and the Monte Carlo with our control variate, namely $VAR_{MC}/VAR^{CV}(P_{BS})$. Comparing the third and fourth columns in Table 6, a significant variance reduction is readily observed. For each choice of the pair (α, δ) in Table 6, the computing times of the basic Monte Carlo method, importance sampling method, and control variate method take about 0.3, 3.1, 2.4 seconds, sampling method, and control variate method take about 0.3, 3.1, 2.4 seconds, respectively when we implement them in Matlab 6 on a computer equipped with an Intel Centrino 2.2GHz CPU. From these test examples and indeed from other extensive numerical experiments, the control variate given above is superior to the importance sampling.

In [2], a detail account for the accuracy of (martingale) control variate method is analyzed and a comment on the difficulty to analyze the variance associated with importance sampling is stated. For some option price P or its approximation, the martingale term $\mathcal{M}(P, T)$ defined by

$$\mathcal{M}(P,T) = \int_0^T e^{-rs} \frac{\partial P}{\partial x}(s, S_s)\sigma S_s dW_s^{(0)\star}$$

can be interpreted as the delta hedging portfolio accumulated up to time T from time 0. Thus the term $\mathcal{M}(P; T)$ is also called the hedging martingale by the price P and the estimator defined from (3.23), i.e.

$$\frac{1}{N} \sum_{i=1}^{N} \left[e^{-rT} H(S_T^{(i)}) - \mathcal{M}^{(i)}(P_{BS}, T) \right], \tag{3.25}$$

is called the martingale control variate estimator. Intuitively the effectiveness of the martingale control variate $e^{-rT}H(S_T) - \mathcal{M}(P_{BS}; T)$ is due to the fact that if delta trading $\frac{\partial \tilde{P}}{\partial x}(t, x)$ is closed to the actual hedging strategy, fluctuations of the replicating error will be small so that the variance of the estimator (3.25) should be small. Under OU-type processes to model (Y_t, Z_t) in (2.1) with $0 < \varepsilon, \delta \ll 1$, the variance of the martingale control variate for European options is small of order ε and δ. This asymptotic result is shown in [2]. Variance analysis to American options and Asian options can be found in [3] and [8] respectively.

4. Quasi Monte Carlo Method and a Counterexample

All Monte Carlo methods studied so far are fundamentally related to pseudo random sequences that generate random samples in our simulations. As an alternative integration methods, the use of the quasi-random sequences (also called low-discrepancy sequences) to generate random samples needed in simulations is called Quasi-Monte Carlo method. QMC method has drawn a lot of attention in financial applications, for example see [12] and [20], because they are able to provide better convergence rates.

4.1 Introduction to Quasi Monte Carlo Method

There are two classes of low-discrepancy sequences (LDS in short) as explained extensively in [10], [16] and [23]. One is called the digital net sequences, such as Halton sequence, Sobol sequence, Faure sequence, and Niederreiter (t, s)–sequence, etc. To estimate an integral with a smooth integrand over a hypercube space, this kind of LDS has convergence rate $O(\frac{(\log N)^s}{N})$, where s

Table 7. Comparisons of low-biased estimates (Column 2–7) and the actual American option prices (Column 8). MC denotes the basic Monte Carlo estimates. MC+CV denotes the control variate estimates with the hedging martingale $\mathcal{M}(P_E; \underline{\tau})$ being the additive control. Standard errors are shown in the parenthesis. QMC and QMC+CV denote calculations of Equation (4.27) and (4.28) using quasi sequences respectively.

S_0	MC	MC+CV	Nied	Nied+CV	Sobol	Sobol+CV	P_0(true)
80	20.7368 (0.2353)	20.6876 (0.0124)	22.6082	**21.3201**	21.7576	**21.6013**	21.6059
85	17.3596 (0.2244)	17.3586 (0.0134)	19.4384	**17.9261**	18.3297	**18.0350**	18.0374
90	14.3871 (0.2125)	14.3930 (0.0139)	16.6127	**14.8104**	15.1183	**14.8961**	14.9187
95	11.8719 (0.1995)	11.8434 (0.0148)	13.8932	**12.1300**	12.3594	**12.1673**	12.2314
100	9.8529 (0.1881)	9.6898 (0.0157)	11.6586	**9.8260**	10.0792	**9.8754**	9.946
105	7.9586 (0.1684)	7.8029 (0.0154)	9.6797	**7.8798**	8.1467	**7.9813**	8.0281
110	6.2166 (0.1518)	6.2606 (0.0150)	7.9423	**6.2726**	6.4757	**6.3924**	6.4352
115	5.0815 (0.1367)	5.0081 (0.0144)	6.5097	**4.9778**	5.1302	**5.0681**	5.1265
120	4.0885 (0.1245)	3.9389 (0.0146)	5.3008	**3.8998**	3.9772	**3.9784**	4.0611

is the dimension of the problem and N denotes the number of quasi-random sequence. The other class is the integration lattice rule points. This type of LDS is especially efficient for estimating multivariate integrals with periodic and smooth integrands, and it has convergence rate $O(\frac{(\log N)^{\alpha s}}{N^{\alpha}})$, where $\alpha > 1$ is a parameter related to the smoothness of the integrand. L'Ecuyer [14] also made contributions to lattice rules based on linear congruential generator. One important feature of this type of lattice rule points (referred to L'Ecuyer's type lattice rule points, LTLRP, thereafter) is that it is easy to generate high dimensional LTLRP point sets with convergence rate comparable to digital net sequences. We will apply the LTLRP as well since our test examples are high dimensional. Besides the above LDS, we also apply the Brownian bridge (BB) sampling technique to our test problems. Detailed information about Brownian bridge sampling can be found in [7].

It is possible to measure the QMC error through a confidence interval while preserving much of the accuracy of the QMC method. Owen [19] showed that for smooth integrands, the root mean square error of the integration over the hypercube space using a class of randomize nets is $O\left(1/N^{1.5-\varepsilon}\right)$ for all $\varepsilon > 0$. This accuracy result promotes the use of *randomized* QMC methods. See for example [11] and [14] for the use of randomization schemes.

Despite that regularity of the integrand function corresponding to the payoff $H(S_T)$ is generally poor [7], there are still many applications of using QMC or randomized QMC as a computational tool. In Section 4.2 we give a counterexample of using QMC method for pricing lower bound solutions of American put options. The error of QMC methods applying to the basic Monte Carlo estimator can be very sensitive to the choice of quasi random sequences.

As shown in Table 7, low-biased prices calculated from either Niederreiter or Sobol sequence are found greater than the benchmark true American option prices in all cases of intial stock price $S(0)$, though Sobol sequence does generate smaller estimates than Niederreiter. Surprisingly by adding a hedging martingale as a control to construct the new estimator of control variate, the accuracy of the low-biased America option price estimates are found to be significantly better for
(1) Monte Carlo Simulations shown on Columns 2 and 3 in Table 7
(2) two quasi-random sequences shown on Columns 4, 5 (Niererreiter) and 6, 7(Sobol) respectively.

In particular we observe that the low-biased estimates obtained from Sobol sequence corrected by martingale control (Sobol +CV shown on Column 7) are even more accurate than estimates obtained from martingale control variate (MC + CV) shown on Column 3. This effect documents that the hedging martingale improve the smoothness of the original American option pricing problem for QMC method.

4.2 Low-Biased Estimate of American Put Option Price

The right to early exercise a contingent claim is an important feature for derivative trading. An American option offers its holder, not the seller, the right but not the obligation to exercise the contract any time prior to maturity during its contract life time. Based on the no arbitrage argument, the American option price at time 0, denoted by P_0, with maturity $T < \infty$ is considered as an optimal stopping time problem [3, 22] defined by

$$P_0 = \sup_{0 \le \tau \le T} \boldsymbol{E}^\star \left\{ e^{-r\tau} H(S_\tau) \right\}, \tag{4.26}$$

where τ denotes a bounded stopping time less than or equal to the maturity T. We shall assume in this section that the underlying dynamics S_t follows Black-Scholes model so that $dS_t = rS_t dt + \sigma S_t dW_t^\star$.

Longstaff and Schwartz [17] took a dynamic programming approach and proposed a least-square regression to estimate the *continuation value* at each in-the-money asset price state. By comparing the continuation value and the instant exercise payoff, their method exploits a decision rule, denoted by $\underline{\tau}$, for early exercise along each sample path generated. As the fact that $\underline{\tau}$ being a suboptimal stopping rule, Longstaff-Schwartz' method induces a low-biased American option price estimate

$$\boldsymbol{E}^\star \left\{ e^{-r\underline{\tau}} H\left(S_{\underline{\tau}}\right) \right\}. \tag{4.27}$$

It is shown in [2] and [3] that we can use a locally hedging martingale to preserve the low-biased estimate (4.27) by

$$\mathbf{E}^{\star}\left\{e^{-r\underline{\tau}}H\left(S_{\underline{\tau}}\right)-\int_{0}^{\underline{\tau}}e^{-rs}\frac{\partial P_E}{\partial x}(s,S_s)\sigma S_s dW_s^{\star}\right\} \tag{4.28}$$

where $\tilde{P}$ is an approximation of the American option price. By the spirit of hedging martingale discussed in Section 3.2, we consider $\tilde{P} = P_E$ the counterpart European option price. In the case of the American put option, P_0 is unknown but its approximation P_E admits a closed-form solution, known as the Black-Scholes formula. Its delta, used in (4.28), is given by

$$\frac{\partial P_E}{\partial x}(t,x;T,K,r,\sigma) = \mathcal{N}\left(\frac{\ln(x/K)+(r+\sigma^2/2)(T-t)}{\sigma\sqrt{T-t}}\right)-1,$$

where $\mathcal{N}(x)$ denotes the cumulative normal integral function.

As an example we consider a pricing problem at time 0 for the American put option with parameters $K = 100, r = 0.06, T = 0.5$, and $\sigma = 0.4$. Numerical results of the low-biased estimates by MC/QMC with or without hedging martingales are demonstrated in Table 7. The first column illustrates a set of different initial asset price S_0. The true American option prices corresponding to S_0 are given in Column 8, depicted from from Table 1 of [22]. Monte Carlo simulations are implemented by sample size $N = 5000$ and time step size (Euler discretization) $\Delta t = 0.01$. Column 2 and Column 3 illustrate low-biased estimates and their standard errors (in parenthesis) obtained from MC estimator related to Equation (4.27) and MC+CV estimator related to Equation (4.28) respectively.

For Monte Carlo method, we observe that (1) almost all estimates obtained from martingale control variate are below the true prices (2) the standard errors are significantly reduced after adding the martingale control $\mathcal{M}(P_E;\underline{\tau})$. A variance analysis for the applications of Monte Carlo methods to estimate high and low-biased American option prices can be found in [3].

For QMC methods we use 5000 Niederreiter and Sobol sequences of dimension 100. In column 4 we see clearly that in all cases of S_0, low-biased QMC estimates are unreasonably greater than the true American prices. The striking part is that after adding hedging martingales, low-biased estimates related to Nied+CV shown on Column 5 and Sobol+CV shown on Column 7 are indeed below the benchmark true prices. These results strongly indicate that the hedging martingale plays the role of a smoother for MC/QMC methods. Because the complexity of American option pricing problems is high, we explore the smooth effect of hedging martingales by considering European option pricing problems under multifactor stochastic volatility models in next section.

5. A Smooth Estimator: Control Variate for MC/QMC Methods

We have seen in Section 3 the (martingale) control variate method performs better in variance reduction than importance sampling for pricing European op-

Table 8. Comparison of simulated European call option values and variance reduction ratios for $\alpha = 50, \delta = 0.5$.

N	MC	MC+CV	Sobol	Sobol+CV	Sobol+BB	Sobol+CV+BB
1024	11.839(0.126)	45.8	5.0	339.3	2.6	129.3
2048	11.837(0.090)	48.0	2.3	304.8	4.0	138.4
4096	11.862(0.064)	48.4	1.8	124.8	2.5	158.4
8192	11.804(0.045)	47.4	2.3	124.0	2.9	148.4
16384	11.816(0.032)	47.1	1.4	176.1	7.7	115.5
32768	11.857(0.022)	48.1	1.7	235.9	4.5	479.9

N	LTLRP	LTLRP+CV	LTLRP+BB	LTLRP+CV+BB
1021	2.0	75.5	7.3	687.9
2039	3.1	135.1	7.0	298.5
4093	3.1	143.9	2.2	140.1
8191	4.2	347.8	4.9	286.0
16381	3.1	227.9	7.8	94.8
32749	6.4	728.7	15.1	741.6

tion prices under multi-factor stochastic volatility models, and in Section 4 that a possibility of misusing QMC method and a way to fix it by (local martingale) control variate for pricing American options. In this section we give a comprehensive studies in efficiencies, including variance reduction ratios and computing time, of using control variate technique developed in the previous section, combined with Monte Carlo and quasi-Monte Carlo methods. Our test examples are European option prices and its *delta*, the first partial derivative of option price with respect to its initial stock price, under random volatility environment.

5.1 European Call Option Estimation

We assume that the underlying asset S is given by (2.1). In our computations, we use C++ on Unix as our programming language. The pseudo random number generator we used is *ran*2() in [21]. In our comparisons, the sample sizes for MC method are 10240, 20480, 40960, 81920, 163840, and 327680, respectively; and those for Sobol sequence related methods are 1024, 2048, 4096, 8192, 16384, and 32768, respectively, each with 10 random shifts; and the sample sizes for L'Ecuyer's type lattice rule points (LTLRP for short) related methods are 1021, 2039, 4093, 8191, 16381, and 32749, respectively, and again, each with 10 random shifts.

In the following examples, we divide the time interval $[0, T]$ into $m = 128$ subintervals. In Table 8, the first column labeled as N indicates the number of Monte Carlo simulations or the Quasi-Monte Carlo points. The second column labeled as MC indicates the option price estimates (standard errors in the parenthesis) based on the basic MC estimator (3.12). All rest columns record variance reduction ratios between many specific MC/QMC methods and the basic MC estimates. For example, the third column labeled as MC+CV indicates the variance

Table 9. Comparison of time (in seconds) used in the simulation of the above European option.

N	MC	MC+CV	Sobol	Sobol+CV	Sobol+BB	Sobol+CV+BB
1024	7	10	7	9	7	12
2048	13	19	13	17	13	18
4096	26	40	26	35	28	39
8192	54	82	56	70	56	78
16384	109	167	107	139	107	157
32768	225	316	222	301	218	318

N	LGLP	LGLP+CV	LGLP+BB	LGLP+CV+BB
1021	5	9	7	11
2039	11	17	13	21
4093	24	34	27	40
8191	45	72	56	78
16381	90	167	106	160
32749	184	293	219	311

reduction ratios as the squares of the standard errors in the second column versus the standard errors obtained from the martingale control variate estimation (3.25). The fourth column labeled as Sobol indicates the variance reduction rations as the squares of the standard errors in the second column versus the standard errors obtained from the estimation (3.12) by randomized Sobol sequence.

Model parameters and initial setup of the European call option pricing problems under two-factor stochastic volatility models are chosen the same in Table 4 and Table 5 respectively. For mean-reverting rates α and δ in volatility processes, we take $\alpha = 50$, $\delta = 0.5$. Numerical results are listed in Tables 8 and 9, where MC+CV stands for Monte Carlo method using control variate technique, Sobol+BB means the quasi-Monte Carlo method using Sobol sequence with Brownian bridge sampling technique, LTLRP for QMC method using L'Ecuyer type lattice rule points, etc.

From Table 8, we observed the following facts. Using the control variate technique, the variance reduction ratios are around 48 for pseudo-random sequences. Without control variate, for both Sobol sequence and L'Ecuyer type lattice rule points, even combined with Brownian bridge sampling technique, the variance reduction ratios are only a few times better than the MC sampling. However, when combined with control variate, the variance reduction ratios for the Sobol sequence vary from about 124 to 339 for Sobol+CV and from 115 to 480 for Sobol+CV+BB; and the variance reduction ratios for the L'Ecuyer type lattice rule points range from about 75 to 729 for LTLRP+CV and from 94 to 742 for LTLRP+CV+BB. This implicitly indicates that the new controlled payoff $e^{-rT}(S_T - K)^+ - \mathcal{M}(P_{BS})$ is smoother than the original call payoff $e^{-rT}(S_T - K)^+$. It can be easily seen that under the Black-Scholes model with the constant volatility σ, the controlled payoff is exactly equal to the Black-Scholes option price

$P_{BS}(0, S_0; \sigma)$, which is a constant so as a smooth function; while the original call payoff function is only continuous and even not differentiable.

Another interesting observation is that the variance reduction ratios do not always increase when the two low-discrepancy sequences are combined with control variate and Brownian bridge sampling, compared with when they are combined with control variate without Brownian bridge sampling.

Regarding time used in simulations, from Table 9 we observed that the time differences among methods without control variates are not significant, but the time differences between methods with and without control variates are not ignorable. Similar conclusions are true regarding time used in simulations for other cases. All the programs were run on a main frame using C++ under the UNIX operating system. Note that there were many other jobs were also running when our programs were run on the main frame machine.

5.2 Accuracy Results

To see the smooth effect of a martingale control, Theorem 1 [2] shows that $Var\left(e^{-rT}H(S_T) - \mathcal{M}_0(P_{BS})\right) \leq C\max\{\varepsilon, \delta\}$ for smooth payoff function H when ε and δ are small enough. That is, the original variance $Var\left(e^{-rT}H(S_T)\right)$ is reduced from order 1 to small order of ε and δ using the martingale control. Then by Proinov bound [16] it is easy to show that the error of QMC method is of order $\sqrt{\varepsilon}$ and $\sqrt{\delta}$. Thus it implied that the variance of randomized QMC is of small order ε and δ.

5.3 Delta Estimation

Estimating the sensitivity of option prices over state variables and model parameters are important for risk management. In this section we consider only the partial derivative of option price with respect to the underlying risk asset price, namely *delta*. To compute Delta, we adopt (1) pathwise differentiation (2) central difference approximation to formula our problems. Then as in previous section we use martingale control variate in Monte Carlo simulations and a combination of martingale control variate with Sobol sequence in randomized QMC method.

By pathwise differentiation (see [7] for instance), the chain rule can be applied to Equation (2.2) so that

$$\frac{\partial P^{\varepsilon,\delta}}{\partial S_0}(0, S_0, Y_0, Z_0) = \boldsymbol{E}^\star\left\{e^{-rT}\mathbf{I}_{\{S_T>K\}}\frac{\partial S_T}{\partial S_0} \mid S_0, Y_0, Z_0\right\}$$

is obtained. Since

$$e^{rT}\frac{\partial S_T}{\partial S_0} = e^{\int_0^T \sigma_t dW_t^{(0)\star} - \frac{1}{2}\int_0^T \sigma_t^2 dt} \tag{5.29}$$

is an exponential martingale, one can construct a $\boldsymbol{P}^\star$-equivalent probability measure $\tilde{P}$ by Girsanov Theorem. As a result, under the new measure $\tilde{P}$ the delta

$\frac{\partial P^{\varepsilon,\delta}}{\partial S_0}(0, S_0, Y_0, Z_0)$ has a probabilistic representation of the digital-type option

(5.30) $$P_D^{\varepsilon,\delta}(0, S_0, Y_0, Z_0) := \frac{\partial P^{\varepsilon,\delta}}{\partial S_0}(0, S_0, Y_0, Z_0) = \tilde{E}\left\{\mathbf{I}_{\{S_T > K\}} \mid S_0, Y_0, Z_0\right\},$$

where the dynamics of S_t must follow

(5.31) $$dS_t = \left(r + f^2(Y_t, Z_t)\right) S_t dt + \sigma_t S_t d\tilde{W}_t^{(0)},$$

with $\tilde{W}^{(0)}$ being a standard Brownian motion under $\tilde{P}$. The dynamics of Y_t and Z_t will change according to the drift change of $W_t^{(0)}$.

Following the same argument of option price approximation, or see Appendix in [2], the digital call option $P_D^{\varepsilon,\delta}(0, S_0, Y_0, Z_0)$ admits the *homogenized* approximation $\bar{P}_D(\bar{S}_0, Z_0) := \bar{E}\left\{\mathbf{I}_{\{\bar{S}_T > K\}} \mid \bar{S}_0 = S_0, Z_0\right\}$, where the "homogenized" stock price $\bar{S}_t$ satisfies

$$d\bar{S}_t = \left(r + \bar{\sigma}^2(Z_t)\right) \bar{S}_t dt + \bar{\sigma}(Z_t) \bar{S}_t d\bar{W}_t^{(0)}$$

with $\bar{W}_t^{(0)}$ being a standard Brownian motion [4]. In fact, the homogenized approximation $\bar{E}\left\{\mathbf{I}_{\{\bar{S}_T > K\}} \mid \bar{S}_t, Z_t\right\}$ is a probabilistic representation of the homogenized "delta", $\frac{\partial P_{BS}}{\partial x}$, where P_{BS} defined in Section 2.1.

The martingale control for the digital call option price (5.30) can be constructed as in Section 3.2 so that similar martingale control variate estimator is obtained as

$$\frac{1}{N} \sum_{k=1}^{N} \left[e^{-rT} \mathbf{I}_{\{S_T^{(k)} > K\}} - \mathcal{M}^{(k)}(\bar{P}_D, T) \right].$$

Numerical results of variance reduction by MC/QMC to estimate *delta* can be found in Table 10. All model parameters, initial conditions and mean-reverting rates are chosen the same in previous section.

Another way to approximate the *delta* is by central difference. A small increment $\Delta_S > 0$ is chosen to discretize the partial derivative by

$$P_D^{\varepsilon,\delta} = \frac{\partial P^{\varepsilon,\delta}}{\partial S_0} \approx \frac{P^{\varepsilon,\delta}(0, S_0 + \Delta_S/2, Y_0, Z_0) - P^{\varepsilon,\delta}(0, S_0 - \Delta_S/2, Y_0, Z_0)}{\Delta_S}.$$

Each European option price corresponding to different initial stock price $S_0 + \Delta_S/2$ and $S_0 - \Delta_S/2$ respectively is computed by the martingale control variate method with MC/QMC. Numerical results of variance reduction by MC/QMC to estimate *delta* can be found in Table 11.

In contrast to the European call option cases, QMC method doesn't make a great benefit in variance reduction in both pathwise differentiation and central difference approximation. This is because the regularity of the *delta* function is worse than the call function.

Table 10. Comparison of variance reduction ratios to estimate the Δ of an European call option by martingale control variate method.

N	MC	MC+CV	Sobol	Sobol+CV
1024	0.8320(0.00513)	13.3	2.4	17.5
2048	0.8413(0.00358)	13.2	2.2	8.4
4096	0.8405(0.00254)	13.8	2.9	21.4
8192	0.8371(0.00180)	13.5	3.8	11.3
16384	0.8397(0.00127)	13.7	3.0	21.4
32768	0.8397(0.00090)	13.7	1.3	18.8

Table 11. Comparison of variance reduction ratios to approximate the Δ of an European call option by Central Difference Scheme.

N	MC	MC+CV	Sobol	Sobol+CV
1024	0.8490(0.00507)	15.8	2.6	25.7
2048	0.8354(0.00357)	14.7	1.8	7.9
4096	0.8378(0.00253)	14.7	2.8	21.6
8192	0.8355(0.00179)	14.9	5.5	13.6
16384	0.8381(0.00126)	14.7	4.8	19.4
32768	0.8384(0.00090)	14.7	2.8	15.2

6. Conclusion

Using (randomized) QMC methods for irregular or high dimensional problems in computational finance may not be efficient as shown in pricing American option under Black-Scholes model and European option under multifactor stochastic volatility models, respectively. Based on the delta hedging strategy in trading financial derivatives, the value process of a hedging portfolio is considered as a martingale control in order to reduce the risk (replication error) of traded derivatives. For the martingale control, its role as a smoother for MC/QMC methods becomes clear when significant variance reduction ratios are obtained. An explanation of the effect of the smoother under perturbed volatility models can be found in [8].

References

1. Fouque, J.-P. and Han, C.-H., Variance Reduction for Monte Carlo Methods to Evaluate Option Prices under Multi-factor Stochastic Volatility Models, *Quantitative Finance*, **5**, (2004), 1–10.
2. Fouque, J.-P. and Han, C.-H., A Martingale Control Variate Method for Option Pricing with Stochastic Volatility, *ESAIM Probability & Statistics*, **11**, (2007), 40–54.
3. Fouque, J.-P. and Han, C.-H., Asymmetric Variance Reduction for Pricing American Options, to appear on *Mathematical Modeling and Numerical Methods in Finance*, (2008), Edited by A. Bensoussan, and Q. Zhang.
4. Fouque, J.-P., Papanicolaou, G., and Sircar, R., *Derivatives in Financial Markets with Stochastic Volatility*, (2000), Cambridge University Press.
5. Fouque, J. P., Papanicolaou, G., Sircar, R., and Solna, K., Multiscale Stochastic

Volatility Asymptotics, *SIAM Journal on Multiscale Modeling and Simulation*, **2**, (2003), 22–42.

6. Fouque, J.-P. and Tullie, T., Variance Reduction for Monte Carlo Simulation in a Stochastic Volatility Environment, *Quantitative Finance*, (2002).
7. Glasserman, P., *Monte Carlo Methods in Financial Engineering*, (2003), Springer Verlag.
8. Han, C.-H. and Lai, Y., Generalized Control Variate Methods to Price Asian Options, submitted.
9. Heston, S., A Closed-form solution for options with stochastic volatility with applications to bond and currency options, *Review of Financial Studies*, **6**, (1993), 2.
10. Hua, L. and Wang, Y., *Applications of Number Theory in Numerical Analysis*, (1980), Springer-Verlag.
11. Jackel, P., *Monte Carlo Methods in Finance*, (2002), John Wiley & Sons Ltd..
12. Joy, C., Boyle, P., and Tan, K. S., Quasi-Monte Carlo Methods in Numerical Finance, *Management Science*, **42**, (1996), 926–938.
13. Karatzas, I. and Shreve, S. E., *Brownian Motion and Stochastic Calculus*, 2/e, (2000), Springer.
14. L'Ecuyer, P. and Lemieux, C., Variance Reduction via Lattice Rules, *Management Science*, **46**, (2000), 1214–1235.
15. Lapeyre, B., Pardoux, E., and Sentis, R., *Introduction to Monte Carlo Methods for Transport and Diffusion Equations*, (2003), Oxford University Press.
16. Niederreiter, H., *Random Number Generation and Quasi-Monte Carlo Methods*, (1992), SIAM, Philadelphia.
17. Longstaff, F. and Schwartz, E., Valuing American Options by Simulation: A Simple Least-Squares Approach, *Review of Financial Studies*, **14**, (2001), 113–147.
18. Oksendal, B., *Stochastic Differential Equations*, (1998), Springer.
19. Owen, A. B., Scrambled net variance for integrals of smooth functions, *Annals of Statistics*, **25**, (1997), 1541–1562.
20. Paskov, S. and Traub, J., Faster Valuation of Financial Derivatives, *Journal of Portfolio Management*, **22**, (1995), 113–120.
21. Press, W. H., Teukolsky, S. A., Vetterling, W. T., and Flannery, B. P., *Numerical recipes in C: The Art of Scientific Computing*, (1992), New York : Cambridge University Press.
22. Rogers, L. C. G., Monte Carlo valuation of American Options, *Mathematical Finance*, **12**, (2002), 271–286.
23. Sloan, I. and Joe, S., *Lattice Methods for Multiple Integration*, (1994), Clarendon Press, Oxford.

Estimation of the Local Volatility of Discount Bonds Using Market Quotes for Coupon-Bond Options

Hajime Fujiwara†, Masaaki Kijima‡,* and Katsumasa Nishide§

†Nomura Securities Co.,Ltd., E-mail: fujiwara-0h18@jp.nomura.com
‡Tokyo Metropolitan University/Kyoto University, E-mail: kijima@tmu.ac.jp
§Yokohama National University, E-mail: knishide@ynu.ac.jp

The key issue to price derivatives written on a coupon bond is the volatility structure, such as volatility smiles and skews, of the corresponding discount bonds, since the coupon bond is a portfolio of discount bonds. This paper proposes a method based on Dupire (1994) to estimate the local volatility of discount bonds when only the prices of coupon-bond options are observed in the market. Numerical examples show that our method can construct the local volatility structure of discount bonds that is consistent with the market data.

Key words: Coupon bond options, Dupire's method, local volatility, volatility smile, volatility skew.

1. Introduction

This paper proposes a method based on Dupire (1994) to estimate the local volatility of discount bonds, when only the prices of coupon-bond options are observed in the market. Note that, in order to price derivatives traded in the bond market correctly, we need to model the dynamics of the discount-bond prices, in particular the volatility structure, to be consistent with the observed market prices, since a coupon bond is a portfolio of discount bonds.

It is well known that the volatility smiles and skews are observed not only in the stock options market but also in the bond options market. For example, in the Japanese bond market, implied volatility curves become very steep with respect to the maturity after the introduction of the so-called "Zero-Interest Rate Policy" by the Bank of Japan.[1] Therefore, it is important to develop models that capture

*Corresponding author. Daiwa Securities Group Chair, Graduate School of Economics, Kyoto University, Yoshida-Honmachi, Sakyo-ku, Kyoto, 606-8501, Japan. TEL & FAX:+81-75-753-3511

[1]The Bank of Japan quit the Zero-Interest Rate Policy on July 14, 2006. However, the short rate still stays in the lowest range.

such volatility structures from both theoretical and practical points of view.

The volatility models that deal with the volatility smiles or skews are roughly categorized into three groups. The first group is to use stochastic volatility models. In this approach, the volatility is typically assumed to follow a mean reverting diffusion process. Depending on the parameters of the processes of the underlying asset and the volatility, in particular on the correlation between them, a variety of volatility structures can be generated through this approach. See, e.g., Hull and White (1987) and Heston (1993).

The second group is to add jumps to the underlying asset process. This approach was originally proposed by Merton (1976) and is known as jump-diffusion models. By modeling the jump intensity and the jump size distribution appropriately, jump-diffusion models can generate the volatility smiles. See, e.g., Kou (2003) and references therein for details of jump-diffusion models.

The third group is known as local volatility models. In this approach, the volatility is supposed to be a deterministic function of time to maturity and the price of the underlying asset. This approach was first proposed by Dupire (1994) and becomes popular for stock market practitioners because of its simplicity. Rubinstein (1994) as well as Derman and Kani (1994) provided a binomial tree model that capture the local volatility effect to be consistent with observed market data. Li (2000) extended the binomial tree model and proposed a new algorithm to build the extended tree.

Our method is based on the third approach. As Li (2000) noted, the advantages of this approach are: (1) it is a preference-free approach, (2) all contingent claims are priced based on the *single* model that are consistent with the market data.

In the bond market, only coupon-bond options are traded and volatility skews and smiles are observed. It is a common practice that the price of a coupon-bond option is calculated by Black's formula (1976), where the forward price of the coupon bond is assumed to follow a log-normal process. Hence, it seems straightforward to estimate the local volatility of the forward coupon-bond process based on Dupire's method.

Recall, however, that a coupon bond is a portfolio of discount bonds, and coupon bonds with different maturities are strongly correlated each other. Hence, it is not appropriate to estimate the local volatility structure for each coupon bond separately. To price *all* options written on coupon bonds consistently, we have to specify the processes of discount bonds, i.e. more specifically, the local volatility structure of discount-bond processes. Since there are no discount-bond options traded in the market, we are not able to use Dupire's method to estimate the local volatilities of discount-bond processes directly.

In this paper, we propose a method to extract local volatilities of discount bonds from market quotes for coupon-bond options, and apply our method to the Japanese Government Bond (JGB) market. In the JGB market, only options of coupon-bond futures with short maturities are actively traded. There is no market quote for discount-bond options, whereas coupon-bond options with longer maturities are rarely traded. Therefore, we want to construct a model that requires only a few parameters to be estimated; but still consistent with the market data. For such a market, our method can be a useful tool to estimate the local volatility structure of discount bonds.

This paper is organized as follows. In the next section, we review Dupire's local volatility model to be applied for the estimation of volatility structure of discount bonds. The forward-neutral method is a key tool for this purpose. Section 3 states the method to convert the volatility function of coupon bonds to that of discount bonds. The validity of the method is verified by a simulation study. We then apply our method for options in the JGB market in Section 4. The numerical results show that our method works reasonably well even for the actual market. Section 5 concludes the paper. The calculation of volatility function by Dupire's method (1994) is provided in Appendix A for the reader's convenience, while Appendix B calculates the volatility function of coupon bonds when the implied volatility function is given.

2. The Local Volatility Model for Bond Options

In this section, we introduce the local volatility models for both discount bonds and coupon bonds based on Dupire's method. See Appendix A for details of Dupire's method.

2.1 Discount Bonds

Since we consider the pricing problem of discount-bond options, it is necessary to consider a stochastic interest-rate model. A prominent tool in this setting is the forward-neutral method.

Denote the time-t price of a non-defaultable (government) discount bond with maturity T by $v(t,T)$. For $\tau \geq T$, the T-forward price of the discount bond $v(t,\tau)$ at time t is given by

$$v_T(t,\tau) := \frac{v(t,\tau)}{v(t,T)}, \qquad t < T \leq \tau. \tag{2.1}$$

It is well known that the T-forward price $v_T(t,\tau)$ is a martingale under the T-forward measure Q^T. That is, for some volatility process $\sigma(t)$, the stochastic differential equation (SDE for short) for the T-forward price $v_T(t,\tau)$ is given by

$$\frac{\mathrm{d}v_T(t,\tau)}{v_T(t,\tau)} = \sigma(t)\mathrm{d}W_t^T, \tag{2.2}$$

where W_t^T denotes the standard Brownian motion under Q^T. Moreover, the put option price with maturity T and strike price K written on the discount bond $v(t,\tau)$ is given by

$$p(T,K) = v(0,T)\mathrm{E}^T[\{K - v(T,\tau)\}^+],$$

where the current time is assumed to be 0, $\{x\}^+ = \max\{x, 0\}$ and E^T denotes the expectation operator under the T-forward measure Q^T. Since $v(T,\tau) = v_T(T,\tau)$, we obtain

$$p^T(T,K) := \frac{p(T,K)}{v(0,T)} = \mathrm{E}^T[\{K - v_T(T,\tau)\}^+], \tag{2.3}$$

where $p^T(T,K)$ denotes the current T-forward price of the put option. See, e.g., Kijima (2002) for details of the forward-neutral method.

In order to obtain the local volatility function of discount bonds, we follow the idea of Dupire (1994). That is, consider the SDE (2.2) with a deterministic volatility function $\sigma(t)$ in time t. When the Vasicek model (1977) is assumed for the risk-free interest rate $r(t)$, the volatility function $\sigma(t)$ is given by[2]

$$\sigma(t) = \sigma_T(t,\tau) := \frac{\sigma}{a}\left(\mathrm{e}^{-a(\tau-t)} - \mathrm{e}^{-a(T-t)}\right), \qquad t < T \leq \tau. \tag{2.4}$$

In this setting, the current T-forward prices of European options (call and put options) are given by Black's formula (1976) with constant volatility v, where

$$v^2 = \int_0^T \sigma_T^2(s,\tau)\mathrm{d}s.$$

However, again, the implied volatilities calculated from Black's formula exhibit the volatility smiles or skews in practice.

Suppose that the volatility is not only a function of time t but also a function of the underlying asset $v_T(t,\tau)$. It is assumed that the T-forward price follows the SDE

$$\frac{\mathrm{d}v_T(t,\tau)}{v_T(t,\tau)} = \sigma(t, v_T(t,\tau))\mathrm{d}W_t^T, \qquad t \leq T, \tag{2.5}$$

under the T-forward measure Q^T. The current T-forward price of the put option with maturity T and strike price K is given by (2.3). Hence, following the idea of Dupire (1994) presented in Appendix A, we obtain the local volatility function of the T-forward price $v_T(t,\tau)$ as

$$\sigma^2(T,K) = \frac{2\frac{\partial p^T(T,K)}{\partial T}}{K^2\frac{\partial^2 p^T(T,K)}{\partial K^2}} \tag{2.6}$$

[2]We take the Vasicek model as the basis of the local volatility model, since it seems the simplest model in the stochastic interest-rate setting. Of course, it is possible to assume other volatility functions as the basis. Note that the Hull–White model (1990) produces the same volatility function, since the mean-reverting level is irrelevant to derivative prices. Also, recall that the Hull–White model is a special case of the HJM model (1992) with constant volatility structure. See Inui and Kijima (1998) for details.

in order for the model (2.5) to be consistent with the market prices $p^T(T,K)$. However, this formula is valid only when discount-bond options with all maturities and all strikes are actively traded in the market. Note also that the maturity τ of the underlying discount bond $v(t,\tau)$ is implicit in the formula (2.6); cf. (2.3).

2.2 Coupon Bonds

Next, we consider coupon-bond options. Suppose that a coupon bond pays constant cashflow C_i at time T_i, $i = 1, \cdots, N$. The time-t price of the coupon bond is then given by

$$\Theta(t, T_N) = \sum_{i=1}^{N} C_i v(t, T_i), \qquad t < T < T_i.$$

The T-forward price of the coupon bond at time t is expressed as

$$\Theta_T(t, T_N) := \frac{\Theta(t, T_N)}{v(t, T)} = \sum_{i=1}^{N} C_i v_T(t, T_i). \tag{2.7}$$

This shows that the coupon bond can be considered as a portfolio of discount bonds.

Note that practitioners use Black's formula (1976) for the pricing of coupon-bond options. This means that they assume the SDE (2.2) with a deterministic volatility function $\sigma(t)$ for the T-forward price $\Theta_T(t, T_N)$. Hence, it is natural to assume that $\Theta_T(t, T_N)$ follows the SDE

$$\frac{\mathrm{d}\Theta_T(t, T_N)}{\Theta_T(t, T_N)} = \sigma^{CB}(t, \Theta_T(t, T_N))\mathrm{d}W^T(t), \qquad t \le T, \tag{2.8}$$

where $\sigma^{CB}(T, K)$ is the local volatility of the T-forward price of the coupon bond.

Consider plain-vanilla options (call or put options) written on the coupon bond. We denote the put option price with maturity T and strike price K by $q(T, K)$. Let $q^T(T, K) = q(T, K)/v(0, T)$ be the current T-forward price of the put option. Since

$$q^T(T, K) = \mathrm{E}^T[\{K - \Theta_T(T, T_N)\}^+], \tag{2.9}$$

it is readily seen from Appendix A that $q^T(T, K)$ satisfies the partial differential equation (PDE for short)

$$\frac{\partial q^T(T, K)}{\partial T} = \frac{1}{2} K^2 \sigma^{CB}(T, K)^2 \frac{\partial^2 q^T(T, K)}{\partial K^2}. \tag{2.10}$$

It follows that

$$\sigma^{CB}(T, K) = \left[\frac{2 \frac{\partial q^T(T,K)}{\partial T}}{K^2 \frac{\partial^2 q^T(T,K)}{\partial K^2}} \right]^{1/2}, \tag{2.11}$$

as for the case of discount bonds.

Remark 2.1. Consider call options written on the coupon bond. The price of the call option with maturity T and strike price K is denoted by $c(T,K)$. Let $c^T(T,K) = c(T,K)/v(0,T)$. Since

$$c^T(T,K) = \mathrm{E}^T[\{\Theta_T(T,T_N) - K\}^+] = q^T(T,K) + \Theta_T(0,T_N) - K$$

due to the put-call parity, the T-forward price $c^T(T,K)$ also satisfies the PDE (2.10). Hence, in either case, we can derive the functional form of $\sigma^{CB}(T,K)$ from the observed prices $q^T(T,K)$ of coupon-bond (put and call) options using Equation (2.11).

3. The Proposed Method

In this section, we consider the case that only coupon-bond options are traded in the market, and estimate the local volatility of discount bonds from market quotes for them. For this purpose, we first obtain the local volatility function of coupon bonds, assuming that the prices of coupon-bond options with all maturities and all strikes are observed in the market. The local volatility function is then converted to that of discount bonds under some assumptions on the relationship between them. For notational simplicity, we denote in what follows the T-forward prices $v_T(t,T_i)$ and $\Theta_T(t,T_j)$ by $v_i(t)$ and $\Theta_j(t)$, respectively, where T_i, $T_j > T$.

Suppose, rather than (2.5), that the T-forward price $v_i(t)$ follows the SDE

$$\frac{\mathrm{d}v_i(t)}{v_i(t)} = \sigma_T(t,T_i)\eta_i(t,v_i(t))\mathrm{d}W^T(t) \tag{3.1}$$

under the T-forward measure Q^T, where $\sigma_T(t,\tau)$ is given by (2.4). Note that the volatility structure in the T-forward price $v_i(t) = v_T(t,T_i)$ may depend on the term structure with maturity T_i. In order to take this effect into account, we leave $\sigma_T(t,T_i)$ in the SDE (3.1) and determine the volatility function $\eta_i(t,v)$ from the market data; see the assumption (3.7) below.

On the other hand, we set the local volatility of the coupon bond as $\sigma_j^{CB}(T,K) = \sigma_B\xi_j(T,K)$, where σ_B is a constant and $\xi_j(t,\Theta)$ is a deterministic function that characterizes the local volatility of the T-forward price $\Theta_j(t)$ of coupon bonds. Then, the T-forward price of coupon bond is expressed as

$$\frac{\mathrm{d}\Theta_j(t)}{\Theta_j(t)} = \sigma_B\xi_j(t,\Theta_j(t))\mathrm{d}W^T(t), \qquad t \le T. \tag{3.2}$$

We leave the constant σ_B in (3.2) to match the implied volatility function (4.1) defined later.

3.1 Relation Between Coupon-Bond Options and Discount Bonds

Black's model (1976) treats a coupon bond as a single asset to price a coupon-bond option. In this framework, it is easy to estimate the local volatility of the coupon-bond process from coupon-bond options by using Dupire's method (see Equation (2.11)). However, since a coupon bond is a portfolio of discount bonds, coupon bonds with different maturities are strongly correlated each other. Hence, we cannot estimate the local volatilities of coupon-bond processes independently.

In order to price *all* options on coupon bonds consistently, it is necessary to consider discount bonds as underlying assets. If the option market for discount bonds existed, there were no problem to estimate the local volatilities $\eta_i(t, v)$ of discount-bond processes as given in (3.1). However, in the actual market, only coupon-bond options are traded. Therefore, we have to extract the information about local volatilities $\eta_i(t, v)$ of discount-bond processes from the prices of coupon-bond options.

Suppose that the volatility functions $\xi_j(t, \Theta)$ are estimated by Dupire's method as in (2.11). The local volatilities $\eta_i(t, v)$ are assumed to be expressed as

$$\eta_i(t, x) = f_i\left(\xi_1(t, x), \xi_2(t, x), \ldots, \xi_n(t, x)\right), \tag{3.3}$$

where the functions $f_i(\xi_1, \xi_2, \ldots, \xi_n)$ are to be determined. Suppose further that each function f_i is sufficiently smooth and can be approximated by the N_pth order Taylor expansion

$$\begin{aligned}\eta_i(t, x) \approx \alpha_{i,0} &+ \sum_{k=1}^{n} \alpha_{i,k}\xi_k(t, x) + \sum_{k_1=1}^{n}\sum_{k_2=1}^{n} \alpha_{i,k_1,k_2}\xi_{k_1}(t, x)\xi_{k_2}(t, x) \\ &+ \cdots + \sum_{k_1=1}^{n} \cdots \sum_{k_{N_p}=1}^{n} \alpha_{i,k_1,\cdots,k_{N_p}}\xi_{k_1}(t, x) \times \cdots \times \xi_{k_{N_p}}(t, x).\end{aligned} \tag{3.4}$$

The approximation (3.4) can be seen as an expansion of the local volatility $\eta_i(t, x)$ using functions $\xi_j(t, x)$.

Since we do not know the function forms of f_i, we can consider the expansion coefficients $\{\alpha_{i,0}, \cdots, \alpha_{i,\cdot}\}$ as free parameters to be estimated. That is, the parameters are estimated by minimizing the square relative errors such that

$$\begin{aligned}&\min_{\{\alpha_{i,0},\cdots,\alpha_{i,\cdot}\}} \sum_j \left(\frac{q^T_{obs}(T_j, K_j) - q^T_{model}(T_j, K_j)}{q^T_{model}(T_j, K_j)}\right)^2, \\ &q^T_{model}(T, K) = \mathrm{E}^T\left[f\left(\Theta(T)\right)\right], \\ &\Theta(T) = \sum_{i=1} C_i v_i(T; \alpha_{i,\cdot}), \qquad \alpha_{i,\cdot} = \{\alpha_{i,0}, \cdots, \alpha_{i,n,\ldots,n}\},\end{aligned} \tag{3.5}$$

where f is an option payoff function.

3.2 The Crude Approximation

As mentioned earlier, only options of coupon-bond futures with short maturities are actively traded in the actual bond market. Hence, we want to construct a model that requires only a few parameters to be estimated.

To this end, suppose that the local volatilities $\eta_i(t, v)$ are expressed as

$$\eta_i(t,x) = f_i\left(\xi_N(t,x)\right), \qquad i = 1,2,\ldots,N, \tag{3.6}$$

where the functions $f_i(\xi)$ are to be determined. This assumption can be justified, when the coupon rate is small compared to the principal so that the dominant term of the coupon-bond price $\Theta(t, T_N)$ is the discount-bond price $v(t, T_N)$.

Now, as before, suppose that each function $f_i(\xi)$ admits the approximation

$$\eta_i(t,x) \;\approx\; \alpha_{i,0} + \alpha_{i,1}\xi_N(t,x) + \alpha_{i,2}\xi_N^2(t,x) + \cdots + \alpha_{i,N_p}\xi_N^{N_p}(t,x), \tag{3.7}$$

where N_p is the truncation point of the Taylor expansion. The coefficients $\alpha_{i,k}$ are estimated by minimizing the square errors given by (3.5). We call (3.7) the *crude approximation* of order N_p.

3.3 Numerical Experiments

The crude approximation (3.7) seems too simple to capture the complex volatility structures $\eta_i(t, x)$ of discount-bond processes at first glance. In this subsection, we perform some numerical study to check the accuracy of the approximation (3.7).

Note first that, given the parameter set $\alpha_i = \{\alpha_{i,\cdot}\}$, we need to employ Monte Carlo simulation to calculate the forward price $v_i(t)$ of discount bond with maturity T_i according to (3.1) and (3.7),[3] since closed form solutions for discount bonds are not available for the general volatility structure. The T-forward price $\Theta_N(T)$ of the coupon bond with maturity T_N is then calculated by (2.7).

Now, we perform the following simulation study. Suppose that the current yield curve is flat at 2%. The underlying asset is the 10-year bond with annual coupons of 2%. The maturity of options are all 0.5 year ($T = 0.5$). We set $\sigma_B = 0.2$ in (3.2) and $a = \sigma = 0.01$ in (2.4). Furthermore, we assume that the volatility function for the coupon bond is known and given by

$$\sigma_B\xi(T,K) = 0.2 \times \exp\left(1 - \frac{K}{\Theta_T(0)}\right). \tag{3.8}$$

For the volatility functions $\eta_i(t, x)$ of discount bonds, we apply the crude approximation (3.7). However, if we assume a different parameter set α_i for each discount bond $v_i(t)$, the number of parameters to be estimated may be larger than the available data in practice. In order to avoid this deficiency, we group discount

[3] We apply the Euler approximation to (3.1) in order to generate sample path of $v_i(t)$.

Table 1. Estimated parameter values when $N_g = 1$.

N_p	0	1	2	3	4
SSRE	0.14115	0.0019	0.00009	0.00008	0.00002
α_0	2.842	−0.303	0.893	1.312	0.492
α_1	—	2.904	0.716	0.450	0.876
α_2	—	—	0.992	0.432	1.068
α_3	—	—	—	0.409	0.675
α_4	—	—	—	—	−0.513

SSRE means the sum of square reative errors, and α_i, $i = 0, \ldots, N_p$ are the estimated parameters, where N_p is the truncation point in (3.9). Higher truncation yields a better fit.

Table 2. Estimated parameter values when $N_g = 2$.

N_p	0	1	2	3	4
SSRE	0.14047	0.0022	0.00009	0.00008	0.00002
$\alpha_{1,0}$	4.982	0.3578	0.612	5.000	0.492
$\alpha_{1,1}$	—	−0.002	0.000	0.000	0.876
$\alpha_{1,2}$	—	—	0.000	0.000	1.068
$\alpha_{1,3}$	—	—	—	0.000	0.675
$\alpha_{1,4}$	—	—	—	—	−0.513
$\alpha_{2,0}$	2.842	−0.303	1.011	1.367	0.492
$\alpha_{2,1}$	—	2.902	0.502	0.388	0.876
$\alpha_{2,2}$	—	—	1.089	0.411	1.068
$\alpha_{2,3}$	—	—	—	0.436	0.675
$\alpha_{2,4}$	—	—	—	—	−0.513

The paremeter set α_1 is applied for discount bonds with maturities less than 5 years, while α_2 is for discount bonds with maturities more than or equal to 5 years. SSRE means the sum of square reative errors, and α_{1i} and α_{2i}, $i = 0, \ldots, N_p$ are the estimated parameters, where N_p is the truncation point in (3.9). Higher truncation yields a better fit.

bonds and use the same parameter for each group. That is, our local volatility structure is given by

$$\eta_i(t, x) = \sum_{j=0}^{N_p} \alpha_{i,j} \xi_N^j(t, x), \qquad i = 1, 2, \ldots, N_g < N, \tag{3.9}$$

where N_g is the number of groups. This idea is also used when applied to the JGB market in the next section.

In this setting, we first generate the T-forward price $\Theta(T)$ by Monte Carlo simulation using (3.2) in order to obtain the theoretical prices of put options using (2.9). Note that this is possible since we assume the local volatility function (3.8) for the coupon bonds. In the actual calculation, we compute the put option values with strike prices 80, 81, 82, …, 100.

Next, we apply our method to compute the parameter sets α_i using (3.5). In

Table 3. Put option prices written on the coupon bond when $N_g = 1$.

Strike	Sim	$N_p = 0$	$N_p = 1$	$N_p = 2$	$N_p = 3$	$N_p = 4$
80	0.482	0.408	0.485	0.484	0.482	0.482
81	0.566	0.496	0.567	0.566	0.565	0.566
82	0.660	0.599	0.661	0.661	0.661	0.661
83	0.768	0.717	0.768	0.768	0.769	0.768
84	0.889	0.852	0.888	0.889	0.890	0.890
85	1.025	1.005	1.023	1.024	1.026	1.026
86	1.178	1.177	1.174	1.175	1.178	1.177
87	1.347	1.369	1.342	1.344	1.347	1.346
88	1.535	1.585	1.529	1.531	1.534	1.533
89	1.742	1.822	1.736	1.737	1.741	1.740
90	1.970	2.084	1.964	1.965	1.968	1.967
91	2.219	2.370	2.214	2.215	2.217	2.217
92	2.491	2.681	2.487	2.488	2.489	2.489
93	2.786	3.018	2.785	2.785	2.785	2.786
94	3.106	3.382	3.107	3.107	3.106	3.107
95	3.451	3.771	3.455	3.455	3.452	3.453
96	3.823	4.187	3.830	3.830	3.824	3.826
97	4.221	4.628	4.236	4.231	4.223	4.226
98	4.646	5.097	4.665	4.659	4.648	4.651
99	5.099	5.592	5.122	5.114	5.101	5.105
100	5.580	6.112	5.608	5.599	5.581	5.587

The column "Sim" indicates the theoretical prices obtained by Monte Carlo simulation, while the columns "N_p" show the prices calculated by our method. N_p is the truncation order.

this simulation study, we consider either $N_g = 1$ or $N_g = 2$. More specifically, when $N_g = 2$, we group discount bonds based on the maturity (under or over 5 years), and the paremeter set α_1 is applied for discount bonds with maturities less than 5 years, whereas the paremeter set α_2 is applied for discount bonds with maturities more than or equal to 5 years.

Tables 1 and 2 list the sum of square relative errors (SSRE) and the estimated parameters α_i for $N_g = 1$ and $N_g = 2$, respectively. Of course, a smaller SSRE means a better fit. It is seen that, as the truncation point N_p increases, we obtain a better fit, as expected. Note that $N_p = 0$ means a flat volatility. On the other hand, the case $N_g = 2$ yields no improvement over the case $N_g = 1$. That is, it seems enough to treat all the discount bonds as the same group when estimating the parameters at least for this example.

Tables 3 and 4 show the put option prices written on the coupon bond. The row "Sim" indicates the prices obtained by the Monte Carlo simulation using (2.9). The number of simulation runs is 100,000 and we suppose that these prices are accurate. On the other hand, the rows "N_p" indicate that the prices are calculated by our method. It is observed that even the first order approximation, i.e. $N_p = 1$, yields a reasonable fit (the maximum error is 0.03). When $N_p = 4$, the absolute

Table 4. Put option prices written on the coupon bond when $N_g = 2$.

Strike	Sim	$N_p = 0$	$N_p = 1$	$N_p = 2$	$N_p = 3$	$N_p = 4$
80	0.482	0.408	0.485	0.484	0.482	0.482
81	0.566	0.496	0.567	0.566	0.566	0.565
82	0.660	0.599	0.661	0.661	0.661	0.661
83	0.768	0.717	0.768	0.768	0.768	0.768
84	0.889	0.852	0.888	0.889	0.890	0.890
85	1.025	1.005	1.023	1.024	1.025	1.026
86	1.178	1.177	1.174	1.175	1.177	1.178
87	1.347	1.369	1.342	1.344	1.346	1.347
88	1.535	1.584	1.529	1.530	1.533	1.534
89	1.742	1.822	1.736	1.737	1.740	1.741
90	1.970	2.083	1.963	1.965	1.968	1.968
91	2.219	2.369	2.213	2.215	2.217	2.217
92	2.491	2.681	2.487	2.488	2.489	2.489
93	2.786	3.018	2.786	2.786	2.786	2.785
94	3.106	3.381	3.109	3.108	3.107	3.106
95	3.451	3.771	3.458	3.456	3.453	3.452
96	3.823	4.188	3.834	3.831	3.827	3.824
97	4.221	4.627	4.237	4.232	4.226	4.223
98	4.646	5.096	4.666	4.660	4.651	4.648
99	5.099	5.591	5.124	5.116	5.105	5.101
100	5.580	6.112	5.610	5.600	5.587	5.582

The column "Sim" indicates the theoretical prices obtained by Monte Carlo simulation, while the columns "N_p" show the prices calculated by our method. N_p is the truncation order.

errors are less than 0.007. Hence, we believe that our method is quite useful to estimate the local volatility of discount bonds as far as the volatility function of coupon bonds is estimated correctly.

4. Application to the JGB Market

In this section, we apply the crude approximation (3.9) to estimate the local volatility functions of discount bonds for the actual market, namely for the JGB market. As explained in the introductory section, the JGB market has a number of special features that make it difficult to determine the accurate volatility model.[4] In the authors' best knowledge, this is the first paper that analyzes the local volatility structure of the JGB market.

4.1 Market Data

In the JGB market, options written on the 10-year bond futures are actively traded and the volatility skews are observed. We provide the market quotes for the

[4]The JGB yield curve exhibits the so-called S shape. Recently, many attempts have been made to construct term structure models that can capture the special shape of the yield curve. See Kabanov, Kijima and Rinaz (2007) and references therein for details.

Table 5. Market quotes for the put options on September 10, 2008.

Option 1		Option 2	
Time to Maturity: 0.139 year		Time to Maturity: 0.216 year	
Strike	IV	Strike	IV
134.5	6.01 %	129	7.32 %
135	5.90 %	130	6.57 %
135.5	5.78 %	131	6.07 %
136	5.64 %	132	5.91 %
136.5	5.54 %	133	5.63 %
137	5.53 %	134	5.55 %
137.5	5.45 %	135	5.2 %
138	5.34 %	136	5.00 %
138.5	4.04 %	137	4.77 %
139	3.66 %	138	4.38 %
139.5	2.96 %	139	4.05 %
		140	3.35 %

Market quotes for the put options written on the 10-year bond futures on September 10, 2008. The implied volatilities are presented in column IV. Note the significant volatility skews.

Table 6. Details of JGB # 274 on September 10, 2008.

Maturity date	December 20, 2015
Time to Maturity	7.28 year
Coupon rate	1.5 %
Conversion Factor	0.745838
Market price	102.44

This JGB is the cheapest deliverable bond, and so we can think that the bond options traded in the market are written on this bond.

put options with times to maturity 0.139 year and 0.216 year on the 10-year bond futures on September 10, 2008 in Table 5. For notational convenience, we express the options with time to maturity 0.139 year as "Option 1" and the options with time to maturity 0.216 year as "Option 2". In either case, a significant volatility skew is observed.

The cheapest deliverable bond of the underlying 10-year bond futures in Table 5 is the JGB 10-year bond series # 274 (JGB # 274 for short). Thus, the options written on the 10-year bond futures are considered as the options written on the JGB # 274 by using the conversion factor. We provide the details of the JGB # 274 in Table 6.

4.2 Implied Volatilities of the Coupon-Bond Options

In order to estimate the volatility function $\xi(t, \Theta)$ of the coupon bond, option prices for all strikes and all maturities are required. To avoid this unrealistic requirement, we first estimate the implied volatility function using the observed option prices and then, based on Black's formula (1976), calculate the prices of

Table 7. Estimated parameters of the implied volatility function.

	Option 1	Option 2
a_1	−18.98	−8.89
a_2	−122.91	−22.28
σ_B	0.0508	0.0441

Table 8. Fitting results of the estimated volatilities.

Option 1			Option 2		
Strike	IV	Model	Strike	IV	Model
134.5	6.01 %	6.03 %	129	7.32 %	6.99 %
135	5.98 %	6.00 %	130	6.57 %	6.68 %
135.5	5.78 %	5.98 %	131	6.07 %	6.37 %
136	5.64 %	5.84 %	132	5.91 %	6.06 %
136.5	5.54 %	5.64 %	133	5.63 %	5.75 %
137	5.53 %	5.36 %	134	5.55 %	5.44 %
137.5	5.45 %	5.02 %	135	5.20 %	5.14 %
138	5.34 %	4.64 %	136	5.00 %	4.84 %
138.5	4.04 %	4.24 %	137	4.77 %	4.55 %
139	3.66 %	3.82 %	138	4.38 %	4.26 %
139.5	2.96 %	3.38 %	139	4.05 %	3.99 %
			140	3.35 %	3.72 %

The column "IV" indicates the observed implied volatility, while the column "Model" represents the volatility calculated from (4.1) using the parameter values listed in Table 7.

coupon-bond options for all maturities and all strikes consistent with the observed implied volatility. The volatility function of coupon bonds can then be determined.

Suppose that the functional form of the implied volatility is given by[5]

$$\sigma_{mkt}(T,K) = \sigma_B \exp\left(\sum_{n=1}^{M} a_n \left(\frac{K}{K_{ATM}(T)} - 1\right)^n\right), \tag{4.1}$$

where $K_{ATM}(T)$ is the *at-the-money* option strike with maturity T. We estimate the parameters σ_B, a_n by the least square fitting to the observed implied volatilities. Because we have only a few options traded in the JGB market, we take $M = 2$ to avoid the overfitting problem. That is, the number of parameters to be estimated is 3 for 11 data in Option 1 and 12 data in Option 2. The estimated parameters are listed in Table 7, while the fitting results are shown in Table 8. The volatility function of the coupon bond calculated from the implied volatility model (4.1) is given in Appendix B.

[5] See Fengler (2006) for a variety of estimation methods of implied volatility.

Table 9. SSRE and estimated parameters α_i for $N_g = 1$.

N_p	0	1	2	3	4
SSRE	2.043	0.171	0.147	0.141	0.132
α_0	0.926	−0.260	0.276	0.492	0.601
α_1	—	0.890	0.082	0.008	0.001
α_2	—	—	0.277	0.041	0.007
α_3	—	—	—	0.112	0.02
α_4	—	—	—	—	0.49

SSRE means the sum of square reative errors, and N_p is the truncation order. A higher truncation order yields a better fit.

Table 10. SSRE and estimated parameters α_i for $N_g = 2$.

N_p	0	1	2	3	4
SSRE	2.014	0.168	0.135	0.135	0.133
$\alpha_{1,0}$	4.451	5.022	5.614	4.992	4.992
$\alpha_{1,1}$	—	0.007	0.001	0.003	0.000
$\alpha_{1,2}$	—	—	0.001	0.000	0.000
$\alpha_{1,3}$	—	—	—	0.000	0.000
$\alpha_{1,4}$	—	—	—	—	0.000
$\alpha_{2,0}$	0.925	−0.253	0.481	0.752	0.573
$\alpha_{2,1}$	—	0.884	−0.224	−0.194	−0.114
$\alpha_{2,2}$	—	—	0.381	−0.0832	0.066
$\alpha_{2,3}$	—	—	—	0.193	0.135
$\alpha_{2,4}$	—	—	—	—	−0.008

The fitting results are similar to the case of $N_g = 1$. Hence, all the discount bonds can be treated as the same group.

4.3 Local Volatility Function of Discount Bonds

Once we obtain the volatility function $\xi(t, \Theta)$ of the coupon bond, we can estimate the local volatility of discount bonds using the procedure presented in the previous section.

For the JGB # 274, the number of coupon payments is 15 ($N = 15$). We use $a = 0.01$ and $\sigma = 0.00973$ for $\sigma_T(t, \tau)$ in (2.4). These values are obtained by fitting the swaption volatilities using the Hull–White model (1990). Moreover, we use the implied volatility parameters (a_1, a_2, σ_B) of Option 1 for $t < 0.139$ and those of Option 2 for $t > 0.139$. As before, when $N_g = 2$, we apply the parameter set α_1 for discount bonds whose maturity is less than 3.5 years and the parameter set α_2 for those with more than or equal to 3.5 years.

The estimated parameters α_i as well as SSRE (sum of square relative errors) are presented in Tables 9 and 10 for the cases of $N_g = 1$ and $N_g = 2$, respectively. It is explicitly observed that a higher order truncation yields a better fit, although the improvement is not as significant as in the previous case (see Tables 1 and 2), where the yield curve was assumed to be flat. Also, the grouping based on the

Table 11. Option prices for $N_g = 1$.

Option 1						
Strike	Market	$N_p = 0$	$N_p = 1$	$N_p = 2$	$N_p = 3$	$N_p = 4$
134.5	0.274	0.258	0.256	0.237	0.248	0.238
135	0.375	0.352	0.343	0.322	0.334	0.323
135.5	0.487	0.469	0.443	0.424	0.434	0.426
136	0.611	0.613	0.557	0.546	0.550	0.547
136.5	0.745	0.789	0.687	0.690	0.684	0.690
137	0.895	0.996	0.836	0.863	0.840	0.860
137.5	1.066	1.236	1.008	1.071	1.025	1.065
138	1.267	1.513	1.215	1.326	1.250	1.314
138.5	1.511	1.825	1.470	1.631	1.527	1.614
139	1.810	2.171	1.795	1.983	1.867	1.961
139.5	2.174	2.546	2.216	2.376	2.266	2.353
Option 2						
Strike	Market	$N_p = 0$	$N_p = 1$	$N_p = 2$	$N_p = 3$	$N_p = 4$
129	0.041	0.008	0.033	0.038	0.035	0.038
130	0.059	0.020	0.056	0.060	0.058	0.060
131	0.085	0.046	0.091	0.090	0.091	0.091
132	0.125	0.092	0.139	0.133	0.137	0.134
133	0.186	0.172	0.206	0.192	0.201	0.192
134	0.277	0.303	0.302	0.281	0.294	0.282
135	0.415	0.501	0.435	0.419	0.428	0.421
136	0.623	0.787	0.621	0.631	0.622	0.633
137	0.928	1.174	0.875	0.951	0.896	0.948
138	1.362	1.671	1.230	1.406	1.290	1.391
139	1.948	2.280	1.744	2.008	1.845	1.984
140	2.691	2.993	2.513	2.752	2.596	2.727

The option prices calculated by Monte Carlo simulation using the estimated volatility functions of discount bonds. It is observed that the differences between the observed prices and the estimated prices are sufficiently small.

maturity makes no difference as far as concerned with the fitting.

On the other hand, Tables 11 and 12 present the option prices calculated by the Monte Carlo simulation using the estimated volatility functions of discount bonds. The absolute differences between the observed prices and the estimated prices are less than 0.20 for the first order case ($N_p = 1$) and 0.18 for the fourth order case ($N_p = 4$).

When the absolute differences are used, the fitness becomes better for at-the-money options (with large premium), while it becomes poor for the out-of-the-money options. The skews are observed for the out-of-the-money options. The reason for this may be that we used the relative error in option prices in the fitting procedure (3.5).

5. Conclusion

In this paper, we proposed a methodology to estimate the local volatility of discount bonds, when only the prices of coupon-bond options are observed in the market. Note that, in order to price derivatives traded in the bond market cor-

Table 12. Option prices for $N_g = 2$.

Option 1						
Strike	Market	$N_p = 0$	$N_p = 1$	$N_p = 2$	$N_p = 3$	$N_p = 4$
134.5	0.274	0.258	0.256	0.245	0.242	0.241
135	0.375	0.351	0.343	0.330	0.327	0.326
135.5	0.487	0.468	0.444	0.431	0.429	0.428
136	0.611	0.613	0.558	0.548	0.547	0.547
136.5	0.745	0.788	0.687	0.684	0.686	0.686
137	0.895	0.995	0.836	0.844	0.850	0.852
137.5	1.066	1.236	1.009	1.035	1.046	1.050
138	1.267	1.512	1.217	1.268	1.285	1.292
138.5	1.511	1.825	1.472	1.555	1.577	1.588
139	1.810	2.170	1.796	1.901	1.923	1.936
139.5	2.174	2.546	2.217	2.298	2.317	2.330
Option 2						
Strike	Market	$N_p = 0$	$N_p = 1$	$N_p = 2$	$N_p = 3$	$N_p = 4$
129	0.041	0.008	0.033	0.036	0.036	0.037
130	0.059	0.020	0.056	0.058	0.059	0.059
131	0.085	0.046	0.091	0.091	0.091	0.091
132	0.125	0.092	0.139	0.137	0.136	0.135
133	0.186	0.172	0.206	0.199	0.197	0.196
134	0.277	0.303	0.302	0.291	0.287	0.287
135	0.415	0.501	0.436	0.427	0.424	0.424
136	0.623	0.786	0.622	0.622	0.626	0.626
137	0.928	1.173	0.875	0.908	0.923	0.927
138	1.362	1.671	1.233	1.319	1.346	1.357
139	1.948	2.280	1.744	1.889	1.926	1.942
140	2.691	2.992	2.514	2.638	2.672	2.687

The option prices calculated by Monte Carlo simulation using the estimated volatility functions of discount bonds. The effect from grouping is not significant.

rectly, we need to model the dynamics of the discount bond prices, in particular the volatility structure, to be consistent with the observed market prices. Recent empirical studies reveal that the effect from the volatility smiles and skews becomes more significant in the bond options market than ever. It is therefore important to develop such models that capture the volatility structure from both theoretical and practical points of view.

Our method consists of two parts, i.e. estimation of the volatility function of coupon bonds using small samples and estimation of the volatility function of discount bonds using the volatility structure of coupon bonds.

As to the first part, given the prices of coupon-bond options, the volatility function of coupon bonds can be determined by Dupire's method (1994). For this purpose, we first estimate the implied volatility from the observed option prices and then, based on Black's formula (1976), calculate the prices of coupon-bond options for all maturities and all strikes consistent with the observed implied volatility.

For the second part, we assume the relationship (3.7) between the volatility functions of coupon bonds and discount bonds. This assumption may be justified when the underlying term structure assumes a single-factor model. However, it is well known that all bond price dynamics of different maturities are not perfectly correlated. Thus, many empirical studies suggest the use of multi-factor term structure models in order to make the model more flexible. Development of multi-factor local volatility models will be our future research.

A. Duprire's Local Volatility Model

As in Dupire (1994), suppose that $S(t)$ follows the following SDE under the risk-neutral measure Q:

$$\frac{\mathrm{d}S(t)}{S(t)} = (r-\delta)\mathrm{d}t + \sigma(t, S(t))\mathrm{d}W_t^Q,$$

where δ is an instantaneous dividend rate.

Let $\phi(T, S)$ be the density function of $S(T)$. From the *forward* Kolmogorov equation, we have

$$\frac{\partial \phi}{\partial T} + \frac{\partial}{\partial S}\{(r-\delta)S\phi(T,S)\} - \frac{1}{2}\frac{\partial^2}{\partial S^2}\{\sigma^2(T,S)S^2\phi(T,S)\} = 0.$$

Twice integrating the above equation with respect to S yields

$$\text{(A.1)} \qquad \frac{1}{2}\sigma^2(T,S)S^2\phi(T,S) = \frac{\partial}{\partial T}\int_0^S \int_0^v \phi(T,u)\mathrm{d}u\mathrm{d}v + (r-\delta)\int_0^S v\phi(T,v)\mathrm{d}v.$$

On the other hand, consider a European put option with maturity T and strike price K, and denote its time-t price by $p(T, K)$. Then, it is readily seen that $p(T, K)$ can be expressed as

$$\text{(A.2)} \qquad p(T,K) = \mathrm{e}^{-r(T-t)}\int_0^K (K-S)\phi(T,S)\mathrm{d}S$$

$$\text{(A.3)} \qquad = \mathrm{e}^{-r(T-t)}\left(K\int_0^K \phi(T,S)\mathrm{d}S - \int_0^K S\phi(T,S)\mathrm{d}S\right).$$

By differentiating (A.2) with respect to K, we get

$$\text{(A.4)} \qquad \frac{\partial p(T,K)}{\partial K} = \mathrm{e}^{-r(T-t)}\int_0^K \phi(T,u)\mathrm{d}u.$$

Therefore, from (A.3), we have an expression for $p(T, K)$ as

$$p(T,K) = K\frac{\partial p(T,K)}{\partial K} - \mathrm{e}^{-r(T-t)}\int_0^K S\phi(T,S)\mathrm{d}S,$$

or equivalently,

$$\text{(A.5)} \qquad \int_0^K v\phi(T,v)\mathrm{d}v = \mathrm{e}^{r(T-t)}\left[K\frac{\partial p(T,K)}{\partial K} - p(T,K)\right].$$

Once again, differentiating (A.4) with respect to K yields

$$\text{(A.6)} \qquad \frac{\partial^2 p(T,K)}{\partial K^2} = \mathrm{e}^{-r(T-t)}\phi(T,K).$$

Equation (A.6) indicates that the price of the European put option satisfies

$$\text{(A.7)} \qquad \int_0^K \int_0^v \phi(T,u)\mathrm{d}u\mathrm{d}v = \int_0^K \int_0^v \mathrm{e}^{r(T-t)}\frac{\partial^2 p(T,u)}{\partial u^2}\mathrm{d}u\mathrm{d}v$$
$$= \mathrm{e}^{r(T-t)}p(T,K),$$

since $\dfrac{\partial p(T,u)}{\partial u} = p(T,u) = 0$ at $u = 0$.

Finally, setting $S = K$ and substituting (A.5)–(A.7) into (A.1), we have

$$\text{(A.8)} \qquad \frac{1}{2}\sigma^2(T,K)K^2\mathrm{e}^{r(T-t)}\frac{\partial^2 p(T,K)}{\partial K^2}$$
$$= \mathrm{e}^{r(T-t)}\left[\frac{\partial p(T,K)}{\partial T} + rp(T,K) + (r-\delta)\left\{K\frac{\partial p(T,K)}{\partial K} - p(T,K)\right\}\right].$$

Solving (A.8) with respect to the volatility function, we get

$$\text{(A.9)} \qquad \sigma^2(T,K) = \frac{2\left[\dfrac{\partial p(T,K)}{\partial T} + \delta p(T,K) + (r-\delta)K\dfrac{\partial p(T,K)}{\partial K}\right]}{K^2\dfrac{\partial^2 p(T,K)}{\partial K^2}}.$$

B. Volatility Function of Coupon Bonds

In this appendix, we derive the volatility function of the coupon bond $\xi(T,\Theta)$ when we assume the implied volatility function (4.1). We demonstrate the calculation for the case of put options only, since the calculation for call options is similar.

Under the implied volatility model (4.1), it follows from (2.9) and Black's formula (1976) that the option price is expressed as

$$q^T(K,T) = KN(\epsilon_K) - \Theta(0)N(\epsilon_K - \sigma_{mkt}(K,T)\sqrt{T}),$$

where

$$\epsilon_K = \frac{\log K - \log\Theta(0)}{\sigma_{mkt}(K,T)\sqrt{T}} + \frac{1}{2}\sigma_{mkt}(K,T)\sqrt{T}.$$

Here, $N(x)$ denotes the cumulative distribution function of the standard normal distribution.

Using the above expression, the first and second derivatives in the right-hand-side of (2.11) are calculated as

$$
\begin{aligned}
\frac{\partial q^T(K,T)}{\partial T} &= KN'(\epsilon_K)\frac{\partial \epsilon_K}{\partial T} - \frac{\partial \Theta(0)}{\partial T} N(\epsilon_K - \sigma_{mkt}(K,T)\sqrt{T}) \\
&\quad - \Theta(0)N'(\epsilon_K - \sigma_{mkt}(K,T)\sqrt{T}) \\
&\quad \times \left(\frac{\partial \epsilon_K}{\partial T} - \frac{\partial \sigma_{mkt}(K,T)}{\partial T}\sqrt{T} - \frac{\sigma_{mkt}(K,T)}{2\sqrt{T}} \right)
\end{aligned} \tag{B.1}
$$

and

$$
\begin{aligned}
\frac{\partial^2 q^T(K,T)}{\partial K^2} &= 2N'(\epsilon_K)\frac{\partial \epsilon_K}{\partial K} + KN''(\epsilon_K)\left(\frac{\partial \epsilon_K}{\partial K}\right)^2 + KN'(\epsilon_K)\frac{\partial^2 \epsilon_K}{\partial K^2} \\
&\quad -\Theta(0)N''(\epsilon_K - \sigma_{mkt}(K,T)\sqrt{T})\left(\frac{\partial \epsilon_K}{\partial K} - \frac{\partial \sigma_{mkt}(K,T)}{\partial K}\sqrt{T}\right)^2 \\
&\quad -\Theta(0)N'(\epsilon_K - \sigma_{mkt}(K,T)\sqrt{T})\left(\frac{\partial^2 \epsilon_K}{\partial K^2} - \frac{\partial^2 \sigma_{mkt}(K,T)}{\partial K^2}\sqrt{T}\right),
\end{aligned} \tag{B.2}
$$

respectively. Here, the first and second derivatives of $N(x)$ are respectively given by

$$
N'(x) = \frac{1}{\sqrt{2\pi}} e^{-\frac{1}{2}x^2}, \qquad N''(x) = \frac{-x}{\sqrt{2\pi}} e^{-\frac{1}{2}x^2}.
$$

The partial derivatives in (B.1) and (B.2) are expressed as

$$
\begin{aligned}
\frac{\partial \epsilon_K}{\partial T} &= -\frac{\frac{\partial \Theta(0)}{\partial T}}{\sigma_{mkt}(K,T)\sqrt{T}\Theta(0)} \\
&\quad -\frac{\log K - \log \Theta(0)}{\sigma_{mkt}^2(K,T)T}\left(\frac{\partial \sigma_{mkt}(K,T)}{\partial T}\sqrt{T} + \frac{\sigma_{mkt}(K,T)}{2\sqrt{T}}\right) \\
&\quad +\frac{1}{2}\frac{\partial \sigma_{mkt}(K,T)}{\partial T}\sqrt{T} + \frac{\sigma_{mkt}(K,T)}{4\sqrt{T}},
\end{aligned} \tag{B.3}
$$

$$
\begin{aligned}
\frac{\partial \epsilon_K}{\partial K} &= \frac{1}{\sigma_{mkt}(K,T)K\sqrt{T}} \\
&\quad -\left[\frac{\log K - \log \Theta(0)}{\sigma_{mkt}^2(K,T)\sqrt{T}} - \frac{\sqrt{T}}{2}\right]\frac{\partial \sigma_{mkt}(K,T)}{\partial K},
\end{aligned} \tag{B.4}
$$

$$
\begin{aligned}
\frac{\partial^2 \epsilon_K}{\partial K^2} = & \frac{-1}{\sigma^2_{mkt}(K,T)K^2\sqrt{T}}\left[\sigma_{mkt}(K,T) + 2K\frac{\partial \sigma_{mkt}(K,T)}{\partial K}\right. \\
& \left. + K^2 \log\left(\frac{K}{\Theta(0)}\right)\frac{\partial^2 \sigma_{mkt}(K,T)}{\partial K^2}\right] \\
& + 2\frac{\log K - \log \Theta(0)}{\sigma^3_{mkt}(K,T)\sqrt{T}}\left(\frac{\partial \sigma_{mkt}(K,T)}{\partial K}\right)^2 + \frac{1}{2}\frac{\partial^2 \sigma_{mkt}(K,T)}{\partial K^2}\sqrt{T}
\end{aligned} \tag{B.5}
$$

and

$$
\frac{\partial \Theta(0)}{\partial T} = R(T)\Theta(0),
$$

where $R(T) = -\frac{\partial \log v(0,T)}{\partial T}$ is the spot rate.

Finally, from (4.1), the derivatives appeared in (B.3)–(B.5) are calculated as

$$
\begin{aligned}
\frac{\partial \sigma_{mkt}(K,T)}{\partial T} &= -\sigma_{mkt}(K,T)\frac{K}{K^2_{ATM}(T)}\frac{\partial K_{ATM}(T)}{\partial T}\sum_{n=1}^{M} a_n n\left(\frac{K}{K_{ATM}(T)} - 1\right)^{n-1}, \\
\frac{\partial \sigma_{mkt}(K,T)}{\partial K} &= \frac{\sigma_{mkt}(K,T)}{K_{ATM}(T)}\sum_{n=1}^{M} a_n n\left(\frac{K}{K_{ATM}(T)} - 1\right)^{n-1}, \\
\frac{\partial^2 \sigma_{mkt}(K,T)}{\partial K^2} &= \frac{\sigma_{mkt}(K,T)}{K^2_{ATM}(T)}\left[\sum_{n=2}^{M} a_n n(n-1)\left(\frac{K}{K_{ATM}(T)} - 1\right)^{n-2}\right. \\
&\quad \left. + \left(\sum_{n=1}^{M} a_n n\left(\frac{K}{K_{ATM}(T)} - 1\right)^{n-1}\right)^2\right],
\end{aligned}
$$

and

$$
K_{ATM}(T) = \frac{\Theta(0, T_N)}{v(0,T)} = \Theta(0).
$$

Thus, we can express $\xi(T, \Theta)$ by substituting (B.1)–(B.5) into (2.11).

References

1. Black, F., The Pricing of Commodity Contracts, *Journal of Financial Economics*, **3**, (1976), 167–179.
2. Black, F. and M. Scholes, The Pricing of Options and Corporate Liabilities, *Journal of Political Economy*, **81**, (1973), 637–654.
3. Brigo, D. and F. Mercurio, *Interest Rate Models: Theory and Practice*, (2001), Springer.
4. Derman, E. and I. Kani, Riding on a Smile, *RISK*, **7**(1), (1994), 32–39.
5. Dupire, B., Pricing with a Smile, *RISK*, **7**, (1994), 18–20.

6. Fengler, M. R., *Semiparametric Modeling of Implied Volatility*, (2006), Springer.
7. Hamida, S. B. and R. Cont, Recovering Volatility from Option Prices by Evolutionary Optimization, *Journal of Computational Finance*, **8**, (2005), 43–76.
8. Heath, D., R. Jarrow, and A. Morton, Bond Pricing and the Term Structure of Interest Rates: A New Methodology for Contingent Claims Valuation, *Econometrica*, **60**, (1992), 77–105.
9. Heston, S., A Closed-Form Solution for Options with Stochastic Volatility with Applications to Bond and Currency Options, *Review of Financial Studies*, **6**, (1993), 327–343.
10. Hull, J. and A. White, The Pricing of Options with Stochastic Volatilities, *Journal of Finance*, **42**, (1987), 281–300.
11. Hull, J. and A. White, Pricing Interest-Rate-Derivative Securities, *Review of Financial Studies*, **3**, (1990), 573–592.
12. Inui, K. and M. Kijima, A Markovian Framework in Multi-factor Heath-Jarrow-Morton Models, *Journal of Financial and Quantitative Analysis*, **33**, (1998), 423–440.
13. Kabanov, Y., M. Kijima, and S. Rinaz, A Positive Interest Rate Model with Sticky Barrier, *Quantitative Finance*, **7**, (2007), 269–284.
14. Kijima, M., *Stochastic Processes with Applications to Finance*, (2002), Chapman & Hall.
15. Kou, S. G., A Jump-Diffusion Model for Option Pricing, *Management Science*, **48**, (2003), 1086–1101.
16. Merton, R., Option Pricing When Underlying Stock Returns Are Discontinuous, *Journal of Financial Economics*, **3**, (1976), 125–144.
17. Li, Y., A New Algorithm for Constructing Implied Binomial Trees: Does the Implied Model Fit Any Volatility Smile?, *Journal of Computational Finance*, **4**, (2000), 69–95.
18. Rebonato, R., *Interest-Rate Option Models*, (1996), Wiley.
19. Rubinstein, M., Implied Binomial Trees, *Journal of Finance*, **49**, (1994), 771–818.
20. Vasicek, O. A., An Equilibrium Characterization of the Term Structure, *Journal of Financial Economics*, **5**, (1977), 177–188.

Real Options in a Duopoly Market with General Volatility Structure

Masaaki Kijima* and Takashi Shibata†

Graduate School of Social Sciences, Tokyo Metropolitan University
*Email: kijima@tmu.ac.jp, †Email: tshibata@tmu.ac.jp

This paper considers strategic entry decisions in a duopoly market when the underlying state variable follows a diffusion with volatility that depends on the current state variable. The extension to this case is more than marginal, since empirical studies have suggested that the volatility is indeed non-constant in real options practices. It is shown that, even in the extended model, three types of equilibria exist in the case of strategic substitution, as for the geometric Brownian case, when the revenue functions are linear. Also, the presence of strategic interactions may push a firm with cost advantage to invest earlier, and the firm value as well as the optimal threshold for the investment decision increases as the market uncertainty increases.

Key words: Investment decision, resolvent operator, volatility, diffusion process, Nash equilibrium, strategic substitution, strategic complement

1. Introduction

The aim of this paper is to analyze the strategies of firms in a duopoly market when the firms have the possibility to make an irreversible investment that increases their profits. The revenue of each firm is uncertain depending on the future condition of the market. By using the option pricing theory, we calculate the opportunity values of the firms and study their strategic entry decisions under the game-theoretic real options approach.

The standard real options approach calculates the value of a project when a monopolistic firm faces uncertainty in the revenue and irreversibility of its investment expenditure. Since we can regard the investment decision making as a financial option, the investment opportunity is similar to an American call option written on the underlying uncertainty. The firm's project value can then be calculated by the option pricing theory in financial engineering.[1] The main result

[1] See, e.g., Epps [6] or Kijima [12] for details of the option pricing theory.

in this framework is that the project value as well as the optimal threshold for the investment decision increases as the market uncertainty increases.[2] An excellent overview of the real options approach is found in Dixit and Pindyck [5] and Trigeorgis [16].

Recently, the basic model has been extended in various ways. Among them, Huisman [10], Pawlina and Kort [14] and others considered the game-theoretic real options approach to incorporate strategic interactions between firms in a duopoly market. These papers focus on strategic aspects associated with timing decisions of firms that operate in an environment with well-specified strategic interactions. In this framework, assuming that the underlying market uncertainty follows a geometric Brownian motion, they showed that there are three types of equilibria in the case of strategic substitution, the presence of strategic interactions may push a firm with cost advantage to invest earlier and, as for the ordinary case, the firm value as well as the optimal threshold for the investment decision increases as the market uncertainty increases (see Huisman [10] and references therein for details).

A geometric Brownian motion is a diffusion process (or a diffusion for short) with constant mean growth rate and constant volatility. In this case, we can obtain the firm value and the optimal threshold for the investment decision in closed form. However, some empirical research suggests that the volatility in real options practices is not constant at all. For example, Davis [4] estimated the volatility when valuing a real option to invest or to abandon and concluded that the volatility varies over the life of the real option. What is of great interest is whether or not the existing results in the game-theoretic real options approach are unchanged for the case of general volatility.

Real option models with general volatilities are not new. For example, Alvarez and Stenbacka [1] considered a multi-stage technology project in which investment opportunities are available at each stage of the project. The multi-stage real option is similar to a *compound* option originally studied by Geske [8]. Assuming a general diffusion process for market uncertainty, they showed that the project value increases as market uncertainty increases, exactly the same result as the standard model (see Alvarez and Stenbacka [1] and references therein for related topics).

In this paper, we extend the results in Alvarez and Stenbacka [1] to the game-theoretic framework. That is, we consider strategic entry decisions in a duopoly market when the underlying state variable follows a diffusion with volatility that depends on the current state variable. Using the resolvent operator, the value functions of each firm for various situations are obtained, i.e. when it is the leader, the

[2] Bernardo and Chowdhry [2] and Shibata [15] considered an *incomplete information* model where firms cannot observe the state variable completely. They showed by numerical experiments that the increased uncertainty may not result in a higher firm value.

follower and when the two firms invest simultaneously. It is shown that the value functions are increasing and convex in the initial state variable and also increasing in market uncertainty. Based on these results and invoking classic results in game theory, we show that the existing results in the game-theoretic real options approach remain unchanged even in our extended setting.

This paper is organized as follows. Section 2 describes the model and provide the information necessary for what follows. In particular, monotonicity and convexity properties of the value functions are discussed in detail. Section 3 obtains the firm values in terms of the resolvent operator for various situations, and discusses the effect of the increased market uncertainty on the optimal thresholds for the investment decisions. We then consider an asymmetric duopoly market in Section 4, i.e. the cases of strategic substitution and strategic complement. Section 5 concludes this paper. Proofs are given in the appendix.

2. The Model and Some Preliminaries

2.1 The Setup

Consider two firms both having the possibility to make an irreversible investment that increases their profits. We denote one firm by i and the other by j, where $i, j \in \{1, 2\}$. The two firms are active on a market to produce a single product and compete with each other to maximize their profits. The revenue of each firm is uncertain depending on the future condition of the product. We describe the uncertainty by state variable X_t which evolves in time as a non-negative stochastic process.

Let $\pi_{N_i N_j}(x)$ denote the revenue flow of firm i when $X_t = x$, $x \geq 0$, where

$$N_k = \begin{cases} 0, & \text{if firm } k \in \{i, j\} \text{ has not invested,} \\ 1, & \text{if firm } k \in \{i, j\} \text{ has invested.} \end{cases}$$

That is, the revenue flow of firm i is equal to $\pi_{11}(x)$ if both firms i and j have invested, $\pi_{10}(x)$ if firm i has invested but firm j has not invested, and so on. It is assumed that $\pi_{N_i N_j}(x)$ is strictly increasing and continuous in x. Hence, higher the state variable $X_t = x$, greater is the profitability of each firm.

As to the relative magnitude relation between the revenue functions, we assume that $\pi_{10}(x) > \pi_{00}(x)$ and $\pi_{11}(x) > \pi_{01}(x)$ for all $x \geq 0$, i.e. the revenue of firm i exceeds its initial revenue, if it invests, regardless of the state of firm j. On the other hand, the relative relation between $\pi_{10}(x)$ and $\pi_{11}(x)$ (also $\pi_{00}(x)$ and $\pi_{01}(x)$) depends on the market condition. Specifically, $\pi_{10}(x) > \pi_{11}(x)$ for all $x \geq 0$ implies that an investment is less profitable when more firms have invested, while $\pi_{10}(x) < \pi_{11}(x)$ implies the opposite situation. The former is called a case of *strategic substitution*, because an increase in supply decreases the revenue. Similarly, the latter is a case of *strategic complement*, because an increase in supply increases the revenue. While the former case is often observed among competing firms in the traditional sense, the latter is plausible when the market is developing.

The relative magnitude relation between the revenue functions is summarized as

$$\text{(RF)}\qquad \begin{array}{ccc} \pi_{10}(x) & > & \pi_{00}(x) \\ \vee \text{ or } \wedge & & \vee \text{ or } \wedge \\ \pi_{11}(x) & > & \pi_{01}(x) \end{array}$$

for all $x \geq 0$. In Section 4, we consider the case of strategic substitution as well as the case of strategic complement.

The sunk cost to adopt the investment opportunity is assumed to be constant and equal to I_i, $i \in \{1, 2\}$, where $I_1 < I_2$. That is, firm 1 has an advantage for the cost relative to firm 2.

In the following, we fix a probability space $(\Omega, \mathcal{F}, P)$ and a filtration $\{\mathcal{F}_t\}_{t \geq 0}$ of sub-σ-algebra satisfying the usual conditions that represents the resolution over time of information commonly available to the firms. It is assumed that $\mathcal{F}_0 = \{\Omega, \emptyset\}$, i.e. it contains no information at time 0.

Each firm seeks an optimal adoption strategy to the investment opportunity. However, since the revenue is uncertain, the adoption time is a random variable. At each time t, the firm must decide whether it should adopt the investment opportunity or not based on the information $\mathcal{F}_t$ available at time t. Mathematically, the adoption time is a $\{\mathcal{F}_t\}$*-stopping time.*

Suppose that the current time is $t \geq 0$, and let τ be the stopping time that firm i adopts the investment opportunity after time t. We denote the set of admissible strategies at time t (i.e. stopping times not less than t) by $\mathcal{T}_t$. The discount factor is constant and equal to $r > 0$. Assuming that the firm is risk neutral, the maximization problem to firm i is given by

$$(1)\qquad V_i(x) = \max_{\tau \in \mathcal{T}_t} E^x \left[\int_t^\tau e^{-r(s-t)} \pi_{0N_j}(X_s) ds + \int_\tau^\infty e^{-r(s-t)} \pi_{1N_j}(X_s) ds - e^{-r(\tau-t)} I_i \middle| \mathcal{F}_t \right],$$

where $E^x[\cdot|\mathcal{F}_t]$ denotes the conditional expectation operator given $\mathcal{F}_t$ with $X_t = x$, $x \geq 0$. The function $V_i(x)$ is called the *value function* of firm i.

We note that the two firms compete each other to maximize their own profits. Hence, the adoption strategy is not only a stopping time, but also it should take the competing firm's strategy into account. In this paper, we seek an optimal adoption strategy to maximize the value function of each firm within the game-theoretic framework. It should be noted that the optimal stopping time may be infinity; in this case, the optimal strategy is never to invest.

Throughout this paper, we call the firm that invests first the *leader* and the other firm the *follower*. The upper-case letters L and F are used to stand for the leader and the follower investment, respectively. The *optimal* stopping time for firm j is denoted by τ_j^k if firm j is k = L, F.

2.2 The Resolvent Operator

In our analyses, we find it useful to define the resolvent operator

(2) $$(R_r\pi)(x) = E^x\left[\int_0^\infty e^{-rs}\pi(X_s)\mathrm{d}s\right], \quad x \geq 0,$$

for a continuous function $\pi(x)$ satisfying the absolute integrability condition, where $E^x[\cdot] = E^x[\cdot|\mathcal{F}_0]$.

In the following, we assume that the underlying state variable $\{X_t\}$ satisfies the stochastic differential equation (SDE)

(3) $$\frac{\mathrm{d}X_t}{X_t} = \mu\mathrm{d}t + \sigma(X_t)\mathrm{d}z_t, \quad t \geq 0,$$

where $\{z_t\}$ denotes a standard Brownian motion. It is assumed that the *mean growth rate* μ is constant and satisfies $\mu < r$, i.e. the mean growth rate of the state variable is strictly less than the discount factor. This together with regularity conditions on the SDE and the function $\pi(x)$ guarantees the existence of the resolvent (2). On the other hand, the *volatility* $\sigma(x)$ is positive for all $x > 0$; but it is any (bounded) function. Recall that, in real options practices, there is some evidence that the volatility is not constant.

Since the volatility $\sigma(x)$ is not constant, the process $\{X_t\}$ satisfying the SDE (3) is no longer a geometric Brownian motion. It is a *diffusion* with linear drift μx and diffusion coefficient $x\sigma(x)$. Recall that the diffusion is a strong Markov process with continuous sample paths. Also, under regularity conditions, $X_t = 0$ implies $X_s = 0$ for all $s > t$, i.e. state 0 is absorbing, because the diffusion coefficient at state 0 is zero.

Before proceeding, we need a couple of preliminary results.

Lemma 2.1. *Suppose that the function $\pi(x)$ is increasing and convex in $x \geq 0$. Then, the resolvent $(R_r\pi)(x)$ is also increasing and convex in x.*

We note that Alvarez and Stenbacka [1] proved the same result when the mean growth rate $\mu(x)$ is convex with a boundedness condition on $(x\sigma(x))'$. In Lemma 2.1, the mean growth rate is constant, while the volatility is arbitrary. It will become clear later why we need to assume the constant mean growth rate in our model.

Lemma 2.2. *Suppose that $\pi(x)$ is linear in x, $\pi(x) = cx + a$ with $c \neq 0$ say. Then, the resolvent $(R_r\pi)(x)$ is also linear and given by*

$$(R_r\pi)(x) = \frac{c}{r-\mu}x + \frac{a}{r}, \quad x \geq 0,$$

provided that $E^x\left[\int_0^\infty (X_s\sigma(X_s))^2\mathrm{d}s\right] < \infty$.

Note that, in the proof of Lemma 2.2, the key assumption is $r > \mu$.

2.3 The Value Functions

In order to discuss the maximization problem (1) with infinite time horizon, we first consider the following *finite* time problem: For any finite $T > 0$ and a function $h(x)$, define

$$V_T(x) = \max_{\tau \in \mathcal{T}} E^x \left[e^{-r \min\{\tau, T\}} h(X_{\min\{\tau, T\}}) \right], \tag{4}$$

where $\mathcal{T} = \mathcal{T}_0$. In the same framework as in the proof of Lemma 2.1, the value function $V_T(x)$ is considered as the price of an American derivative written on the variable $\{X_t\}$ with maturity T, initial state x and payoff function $h(x)$.

Lemma 2.3. *Suppose that $h(x)$ is increasing and convex in $x \geq 0$ with $h(0) = 0$. Then, the value function $V_T(x)$ is also increasing and convex in x with $V_T(0) = 0$.*

We next compare the volatility effect on the value function. Let us denote the diffusion process with volatility $\sigma_i(x)$, $i = 1, 2$, by $\{X_t^i\}$. Accordingly, the value function is denoted by $V_T^i(x)$. The next result states that if $\sigma_1(x)$ dominates $\sigma_2(x)$, i.e. $\sigma_1(x) \geq \sigma_2(x)$ for all $x > 0$, then the value function $V_T^1(x)$ also dominates $V_T^2(x)$ for all $x > 0$. The proof of the next result is similar to Hobson [9] and omitted.

Lemma 2.4. *Suppose that the function $h(x)$ is convex in $x \geq 0$. If $\sigma_1(x) \geq \sigma_2(x)$ for all $x > 0$, then $V_T^1(x) \geq V_T^2(x)$ for all $x > 0$.*

By letting $T \to \infty$, the value function $V_T(x)$ in (4) converges to some function $V_\infty(x)$, under regularity conditions, with the same monotonicity and convexity properties being inherited. That is, suppose that τ is finite with probability one. Under any condition (e.g., uniform integrability) that allows the interchange between the expectation and the limit, one can show that

$$V_\infty(x) = \max_{\tau \in \mathcal{T}} E^x \left[e^{-r\tau} h(X_\tau) \right]. \tag{5}$$

We are now in a position to state potentially useful results for the real options approach.

Proposition 2.1. *Suppose that $h(x)$ is increasing and convex in $x \geq 0$ with $h(0) = 0$. Then, under regularity conditions, the value function $V_\infty(x)$ defined in (5) is also increasing and convex in x with $V_\infty(0) = 0$. The value function is also increasing in volatility in the sense of Lemma 2.4.*

We note that, since the optimization problem given in (5) is of infinite time horizon, under some assumption on h there exists a critical value (optimal threshold) x^* for the state variable X_t such that

$$\tau^* = \inf\{t \geq 0 : X_t = x^*\}, \tag{6}$$

where τ^* denotes the optimal stopping time for (5). See, e.g., Pawlina and Kort [14] and Huisman [10] for details. Since $X_{\tau^*} = x^*$, it follows from (5) that

$$V_\infty(x) = E^x\left[e^{-r\tau^*}\right]h(x^*). \tag{7}$$

Hence, calculation of the value function $V_\infty(x)$ is reduced to obtaining the Laplace transform $E^x[e^{-r\tau^*}]$ of the *first passage time* τ^* to the optimal threshold x^*.

Corollary 2.1. *Suppose that the function $h(x)$ is increasing and convex in $x \geq 0$ with $h(0) = 0$. Then, under regularity conditions, the Laplace transform $E^x[e^{-r\tau^*}]$ of the first passage time τ^* to the optimal threshold x^* is also increasing and convex in x with $E^0[e^{-r\tau^*}] = 0$. The Laplace transform is also increasing in volatility.*

It is well known (or can be easily proved by Dynkin's formula) that the value function (and the Laplace transform) satisfies the following (ordinary) differential equation:

$$\frac{x^2\sigma^2(x)}{2}V_\infty''(x) + \mu x V_\infty'(x) = rV_\infty(x) \tag{8}$$

with boundary conditions

$$V_\infty(0) = 0, \quad V_\infty(x^*) = h(x^*), \quad V_\infty'(x^*) = h'(x^*). \tag{9}$$

Here, the condition $V_\infty(x^*) = h(x^*)$ is called the *value-matching condition*, while $V_\infty'(x^*) = h'(x^*)$ is called the *smooth-pasting condition*.

For example, in the case of a geometric Brownin motion with volatility $\sigma(x) = \sigma$, the value function is given by $V_\infty(x) = Cx^\beta$, where β is a solution to the equation

$$\frac{\sigma^2}{2}\beta(\beta-1) + \mu\beta = r. \tag{10}$$

Note that we have $\beta > 1$ since $V_\infty(0) = 0$ and $r > \mu$. Hence, as obtained in Proposition 2.1, the value function is increasing and convex in x; see Dixit and Pindyck [5] for this result.

The optimal threshold x^* as well as the constant C is obtained by the value-matching condition and the smooth-pasting condition. In the case of a geometric Brownin motion with $h(x) = \max\{Kx - I, 0\}$, where K, $I > 0$, i.e. the ordinary real option model, we need to solve

$$Cx^\beta = Kx - I, \quad C\beta x^{\beta-1} = K,$$

where $x > I/K$. It follows that

$$x^* = \frac{I\beta}{K(\beta-1)}, \quad \beta > 1.$$

Since, from (10), β is decreasing in the volatility σ, the increased market uncertainty results in the increase in the optimal threshold x^* as well as the value function $V_\infty(x^*)$. In what follows, we shall extend these results to a case of a diffusion.

3. A Duopoly Market

In what follows, we assume that the revenue functions are linear.[3] Then, from Lemma 2.2, the resolvent is also linear. Hence, without loss of generality, we assume that $\pi_{N_iN_j}(x) = \pi_{N_iN_j}x$, so that $(R_r\pi_{N_iN_j})(x) = \pi_{N_iN_j}x/(r-\mu)$, $x \geq 0$.

3.1 Follower

Dynamic games are usually solved backwards. We begin by assuming that firm j has already invested as the leader, and shall derive the value function for the follower.

Suppose that the current time is t and firm j has already invested, i.e. $\tau_j^{\mathrm{L}} \leq t$. Then, the value function of firm i as the follower is given by

$$V_i^{\mathrm{F}}(x) = \max_{\tau\in\mathcal{T}_t} E^x\left[\int_t^\tau e^{-r(s-t)}\pi_{01}(X_s)\mathrm{d}s + \int_\tau^\infty e^{-r(s-t)}\pi_{11}(X_s)\mathrm{d}s - e^{-r(\tau-t)}I_i\middle|\mathcal{F}_t\right], \tag{11}$$

where $X_t = x$. Note that, while the first integral in Equation (11) corresponds to the present value (PV) of profits obtained before the investment is undertaken, the second integral represents the PV of profits after the investment. After some algebra, we obtain

$$V_i^{\mathrm{F}}(x) = (R_r\pi_{01})(x) + \max_{\tau\in\mathcal{T}_t} E^x\left[e^{-r(\tau-t)}\left\{(R_r(\pi_{11}-\pi_{01}))(X_\tau) - I_i\right\}\middle|\mathcal{F}_t\right]. \tag{12}$$

For the proof, see the appendix.

Now, as in (6), there exists a point F_i^* such that

$$\tau_i^{\mathrm{F}} = \inf\{s \geq t : X_s = \mathrm{F}_i^*\}.$$

Under our assumptions, the optimal threshold F_i^* is finite and $(\pi_{11}-\pi_{01})\mathrm{F}_i^*/(r-\mu) > I_i$, where $(R_r\pi)(x) = \pi x/(r-\mu)$. It follows that the value function of firm i as the follower is given by

$$V_i^{\mathrm{F}}(x) = \begin{cases} \dfrac{\pi_{11}x}{r-\mu} - I_i, & \text{if } x \geq \mathrm{F}_i^*, \\ \dfrac{\pi_{01}x}{r-\mu} + E^x\left[e^{-r(\tau_i^{\mathrm{F}}-t)}\middle|\mathcal{F}_t\right]\left\{\dfrac{(\pi_{11}-\pi_{01})\mathrm{F}_i^*}{r-\mu} - I_i\right\}, & \text{if } x < \mathrm{F}_i^*. \end{cases} \tag{13}$$

[3]The results with the general volatility structure are essentially due to the linearity of the revenue function.

Therefore, as soon as the value of the state variable X_t reaches the optimal threshold F_i^* (to be determined), firm i should invest immediately as the follower. If the state variable X_t has already exceeded the critical value, there is no need to wait for the investment.

From (13) and Proposition 2.1, we see that the value function of firm i as the follower increases as the market uncertainty increases. As a result, because of the smooth-pasting condition, the optimal threshold F_i^* should also increase in the market uncertainty. We thus have the following.

Proposition 3.1. *Under our assumptions, the value function as well as the optimal threshold for the follower increases as the market uncertainty increases.*

Recall that $I_1 < I_2$ from our earlier assumption, i.e. firm 1 has a cost advantage over firm 2. The next result states that the optimal threshold for firm 1 as the follower comes earlier than firm 2, i.e. $\mathrm{F}_1^* < \mathrm{F}_2^*$.

Proposition 3.2. *Suppose that $I_1 < I_2$. Then, the optimal threshold for firm 1 is always less than that for firm 2.*

3.2 Leader

Suppose that neither firm i nor firm j have invested, and one of them, say j, will be the leader by investing first. In making the calculation, the leader takes into account the strategy of the other firm, say i, after its investment. That is, as soon as the state variable X_t becomes greater than F_i^*, firm i will invest as the follower.

The value of firm j as the leader on the event $\{\tau_j^{\mathrm{L}} \geq t\}$ is given by

$$V_j^{\mathrm{L}}(x) = E^x\left[\int_t^{\tau_j^{\mathrm{L}}} \mathrm{e}^{-r(s-t)}\pi_{00}(X_s)\mathrm{d}s + \int_{\tau_j^{\mathrm{L}}}^{\tau_i^{\mathrm{F}}} \mathrm{e}^{-r(s-t)}\pi_{10}(X_s)\mathrm{d}s \right.$$
$$\left. -\mathrm{e}^{-r(\tau_j^{\mathrm{L}}-t)}I_j + \int_{\tau_i^{\mathrm{F}}}^{\infty} \mathrm{e}^{-r(s-t)}\pi_{11}(X_s)\mathrm{d}s \middle| \mathcal{F}_t\right], \tag{14}$$

where $X_t = x$. The second integral in Equation (14) corresponds to the PV of the leader's profits realized until the follower's investment is made, while the third integral equals the discounted, perpetual stream of profits obtained after the follower's investment. After some algebra, we obtain

$$V_j^{\mathrm{L}}(x) = (R_r\pi_{00})(x) + E^x\left[\mathrm{e}^{-r(\tau_j^{\mathrm{L}}-t)}\left\{(R_r(\pi_{10}-\pi_{00}))(X_{\tau_j^{\mathrm{L}}}) - I_j\right\}\middle|\mathcal{F}_t\right]$$
$$+ E^x\left[\mathrm{e}^{-r(\tau_i^{\mathrm{F}}-t)}(R_r(\pi_{11}-\pi_{10}))(X_{\tau_i^{\mathrm{F}}})\middle|\mathcal{F}_t\right].$$

The proof is similar to that for (12) and omitted.

Let us denote the optimal threshold for τ_j^{L} by L_j^*. Then, similar to (13), we have

$$
(15)\quad V_j^{\mathrm{L}}(x) = \begin{cases} \dfrac{\pi_{11}x}{r-\mu} - I_j, & \text{if } x \geq \mathrm{F}_i^*, \\ \dfrac{\pi_{10}x}{r-\mu} - I_j + E^x\left[\mathrm{e}^{-r(\tau_i^{\mathrm{F}}-t)}\middle|\mathcal{F}_t\right]\dfrac{(\pi_{11}-\pi_{10})\mathrm{F}_i^*}{r-\mu}, & \text{if } \mathrm{L}_j^* \leq x < \mathrm{F}_i^*, \\ \dfrac{\pi_{00}x}{r-\mu} + E^x\left[\mathrm{e}^{-r(\tau_j^{\mathrm{L}}-t)}\middle|\mathcal{F}_t\right]\left\{\dfrac{(\pi_{10}-\pi_{00})\mathrm{L}_j^*}{r-\mu} - I_j\right\} & \\ \quad + E^x\left[\mathrm{e}^{-r(\tau_i^{\mathrm{F}}-t)}\middle|\mathcal{F}_t\right]\dfrac{(\pi_{11}-\pi_{10})\mathrm{F}_i^*}{r-\mu}, & \text{if } x < \mathrm{L}_j^*, \end{cases}
$$

Hence, as soon as the value of the state variable X_t reaches the critical value $x = \mathrm{L}_j^*$ (to be determined), firm j should invest immediately as the leader. If the state variable has already exceeded the critical value, there is no need to wait for the investment as the leader. If $X_t < F_i^*$, then the follower should wait until it reaches the optimal threshold F_i^*; otherwise, there is no need to wait for the investment as the follower.

4. Equilibria

In this section, we investigate the competitive equilibria in our duopoly market. The arguments below are similar to those in Huisman [10] and detailed discussions are largely omitted.

4.1 Strategic Substitution

We first consider the case of strategic substitution, i.e. $\pi_{11} < \pi_{10}$ for all $x \geq 0$. In words, investment is less profitable due to the increase in supply.

Suppose that the current time is $t = 0$, that the leader j has already invested, i.e. $\tau_j^{\mathrm{L}} < 0$, and that the current state variable is below the critical value F_i^*. The last assumption is equivalent to saying that the follower i will not invest now. From the results in the previous sections, we have

$$
(16)\qquad V_j^{\mathrm{L}}(x) = \frac{\pi_{10}x}{r-\mu} - I_j + E^x\left[\mathrm{e}^{-r\tau_i^{\mathrm{F}}}\right]\frac{(\pi_{11}-\pi_{10})\mathrm{F}_i^*}{r-\mu}
$$

and

$$
(17)\qquad V_i^{\mathrm{F}}(x) = \frac{\pi_{01}x}{r-\mu} + E^x\left[\mathrm{e}^{-r\tau_i^{\mathrm{F}}}\right]\left\{\frac{(\pi_{11}-\pi_{01})\mathrm{F}_i^*}{r-\mu} - I_i\right\},
$$

where $\pi_{11} - \pi_{10} < 0$ and $\pi_{11} - \pi_{01} > 0$ by assumptions.

Now, from Corollary 2.1, it is readily seen that the value function $V_i^{\mathrm{F}}(x)$ in (17) is convex in x. Also, it must satisfy the value-matching condition and the smooth-pasting condition, i.e.

$$
(18)\qquad V_i^{\mathrm{F}}(\mathrm{F}_i^*) = \frac{\pi_{11}\mathrm{F}_i^*}{r-\mu} - I_i, \quad \left(V_i^{\mathrm{F}}(\mathrm{F}_i^*)\right)' = \frac{\pi_{11}}{r-\mu},
$$

respectively. The value function $V_i^{\mathrm{F}}(x)$ must satisfy the smooth-pasting condition, because it is determined by optimizing the stopping time τ_i^{F}.

On the other hand, the value function $V_j^{\mathrm{L}}(x)$ in (16) is concave in x and satisfies the value-matching condition

$$V_j^{\mathrm{L}}(\mathrm{F}_i^*) = \frac{\pi_{11}\mathrm{F}_i^*}{r-\mu} - I_j. \tag{19}$$

Note that the function $V_j^{\mathrm{L}}(x)$ does not satisfy the smooth-pasting condition, because it is not optimized in terms of the stopping time τ_i^{F}.

Recalling from Proposition 3.2 that $\mathrm{F}_1^* < \mathrm{F}_2^*$, the value functions $V_1^{\mathrm{F}}(x)$ and $V_1^{\mathrm{L}}(x)$ for firm 1 are depicted in Figure 1, where Fi stands for the critical value F_i^*, $i \in \{1, 2\}$. Note that the curves $V_1^{\mathrm{F}}(x)$ and $V_1^{\mathrm{L}}(x)$ necessarily intersect each other at a point in the interval $(0, \mathrm{F}_1^*)$.

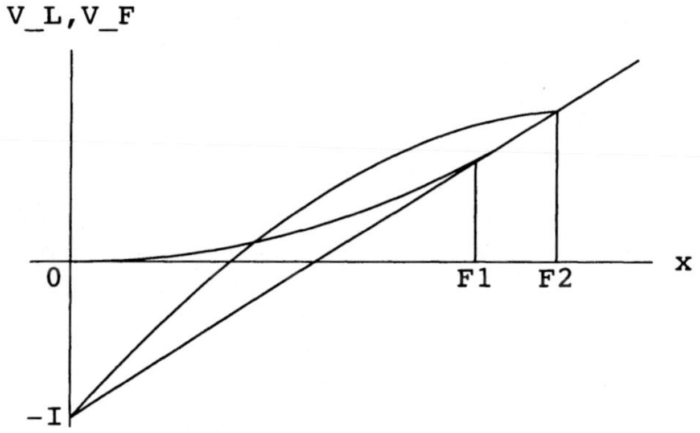

Figure 1. The value functions for firm 1

The story for firm 2 is more involved. In fact, we need to consider the following two cases:

Case (1) $V_2^{\mathrm{L}}(x) < V_2^{\mathrm{F}}(x)$ for all $x \in (0, \mathrm{F}_2^*)$, and

Case (2) $V_2^{\mathrm{L}}(x) \geq V_2^{\mathrm{F}}(x)$ for some $x \in (0, \mathrm{F}_2^*)$.

That is, in Case (2), the follower has an incentive to become the leader when $X_t = x$ such that $V_2^{\mathrm{L}}(x) \geq V_2^{\mathrm{F}}(x)$, while the follower does not have any incentive to become the leader in Case (1). The value functions $V_2^{\mathrm{F}}(x)$ and $V_2^{\mathrm{L}}(x)$ for firm 2 in Case (1) are depicted in Figure 2.

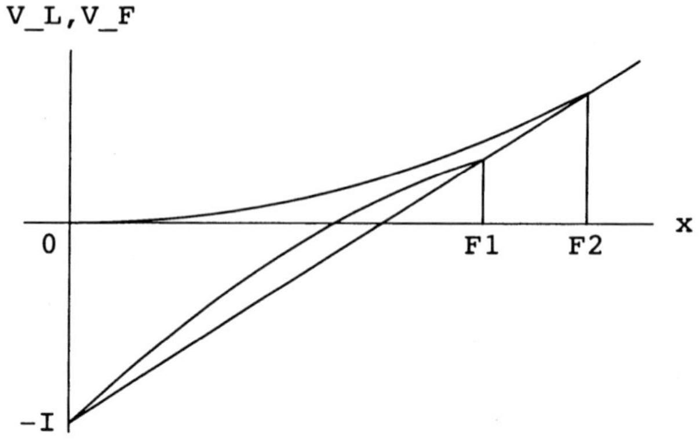

Figure 2. The value functions for firm 2 in Case (1)

Now, since $V_2^{\mathrm{F}}(x)$ is convex and $V_2^{\mathrm{L}}(x)$ is concave in x, the value functions for firm 2 necessarily intersect twice in the interval $(0, \mathrm{F}_1^*)$ for Case (2). Let

$$\mathrm{P}_{21}^* = \inf\{x > 0 : V_2^{\mathrm{L}}(x) \geq V_2^{\mathrm{F}}(x)\}, \quad \mathrm{P}_{22}^* = \sup\{x > 0 : V_2^{\mathrm{L}}(x) \geq V_2^{\mathrm{F}}(x)\}.$$

These points are the crossing points of the value functions $V_2^{\mathrm{F}}(x)$ and $V_2^{\mathrm{L}}(x)$; see Figure 3, where P2i, $i \in \{1, 2\}$, stands for the critical value P_{2i}^*, respectively.

Let L_1^* be the optimal threshold for firm 1 as the leader. The next result is similar to Huisman [10] whose proof is omitted.

Proposition 4.1. *Firm 2 has an incentive to preempt firm 1 on the interval* $[\mathrm{P}_{21}^*, \mathrm{P}_{22}^*]$ *only when* $\mathrm{L}_1^* \in [\mathrm{P}_{21}^*, \mathrm{P}_{22}^*]$.

Firm 2 will try to preempt firm 1 only when firm 1 has not invested before the stopping time τ_{21}^{P}. This may be the situation that the asymmetry among the firms is relatively small and the first-mover advantage is little. Then, the higher-cost firm may have an incentive to preempt the lower-cost firm. This case is called Case (2a). Figure 3 depicts the value functions $V_2^{\mathrm{F}}(x)$ and $V_2^{\mathrm{L}}(x)$ for firm 2 in this case, where L1 stand for the critical value L_1^*. Finally, we note that firm 2 does not preempt firm 1 when $\mathrm{L}_1^* \notin [\mathrm{P}_{21}^*, \mathrm{P}_{22}^*]$. This case is called Case (2b).

Now, we state how the outcome will be derived using the above results.[4] Note that, in Case (2a), there is the possibility to preempt, and we obtain the same outcome as in Huisman [10]. In Cases (1) and (2b), on the other hand, the two firms invest simultaneously.

[4] We use the strategy space and equilibrium concept defined by Huisman [10] and Pawlina and Kort [14]. This concept can be traced back to Fudenberg and Tirole [7].

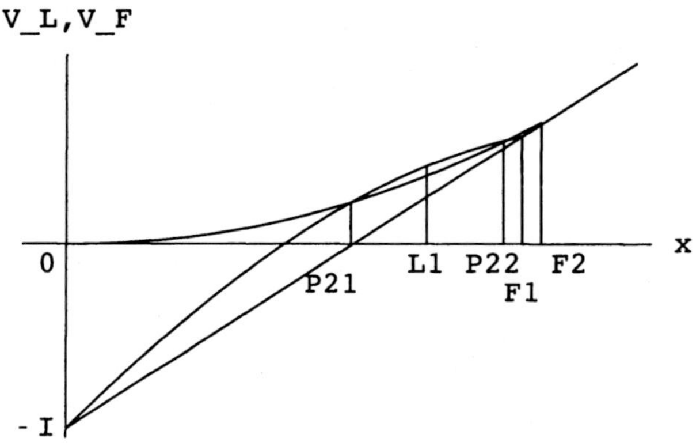

Figure 3. The value functions for firm 2 in Case (2a)

The next result shows that, as for the geometric Brownian case, there are three types of equilibria for the strategic substitution case, when the revenue functions are linear. The proof is exactly the same as in Huisman [10] and omitted. In the next theorem, the current time is $t = 0$ and we let $X_0 = x$.

Theorem 4.1. *If* $L_1^* \in [P_{21}^*, P_{22}^*]$, *i.e. in Case (2a), we have the following for the initial state x:*

(1) If $x \leq P_{21}^*$, *then firm 1 invests at time* τ_{21}^P, *while firm 2 invests at time* τ_2^F.

(2) If $x \in (P_{21}^*, P_{22}^*)$, *then either*

(2-1) firm 1 invests now and firm 2 invests at time τ_2^F,

(2-2) firm 2 invests now and firm 1 invests at time τ_1^F, *or*

(2-3) the two firms invest simultaneounsly now.

(3) If $x \in [P_{22}^*, F_2^*)$, *then firm 1 invests now, while firm 2 invests at time* τ_2^F.

(4) If $x \geq F_2^*$, *then the two firms invest simultanously now.*

If $L_1^* \notin [P_{21}^*, P_{22}^*]$, *i.e. in Cases (1) and (2b), firm 1 invests at time* τ_1^L *and firm 2 invests at time* τ_2^F.

4.2 Strategic Complement

We next consider the case of strategic complement, i.e., $\pi_{11} > \pi_{10}$ for all $x \geq 0$. In this case, investment is more profitable when more firms have invested.

Since $\pi_{11} - \pi_{10} > 0$ and $\pi_{11} - \pi_{01} > 0$, the value functions $V_i^F(x)$ in (17) and $V_j^L(x)$ in (16) are both convex in x for this case. Also, the function $V_i^F(x)$ must satisfy the condition (18), while the function $V_j^{\mathrm{L}}(x)$ satisfies the condition (19) only. Since $\mathrm{F}_1^* < \mathrm{F}_2^*$, we have the complete picture for firm 2, which is depicted in Figure 4.

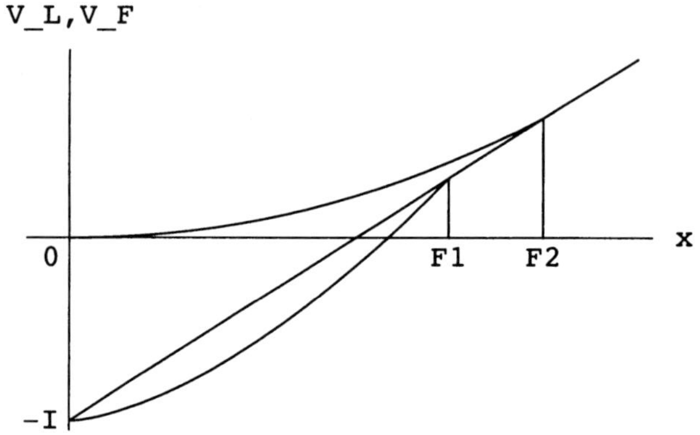

Figure 4. The value functions for firm 2

Different from the strategic substitution case is that the optimal threshold F_i^* is smaller than L_i^*, $i \in \{1, 2\}$, because the simultaneous investment is more profitable than the single investment. Hence, we need to consider the following two cases for firm 1: Case (3) $\mathrm{F}_2^* \geq \mathrm{L}_1^*$, and Case (4) $\mathrm{F}_2^* < \mathrm{L}_1^*$.

In Case (3), it is readily seen that firm 1 invests at time τ_1^{L}, because firm 1 knows that firm 2 is not going to invest before time τ_2^{F}. The value functions $V_1^{\mathrm{L}}(x)$ and $V_1^{\mathrm{F}}(x)$ for firm 1 are depicted in Figure 5 for this case.

On the other hand, in Case (4), the two firms invest simultaneously at time τ_2^{F}, because firm 1 will invest as the leader before time τ_1^{L}; but, firm 1 knows that it is optimal for firm 2 to invest as the follower at time τ_2^{F}. Then, firm 1 initiates the investment and firm 2 will make the simultaneous investment. The value functions $V_1^{\mathrm{L}}(x)$ and $V_1^{\mathrm{F}}(x)$ for firm 1 are depicted in Figure 6 for this case.

Summarizing, we have the following result. The next result shows that, as for the geometric Brownian case, there are two types of equilibria for the strategic complement case, when the revenue functions are linear.

Theorem 4.2. *If* $\mathrm{L}_1^* \leq \mathrm{F}_2^*$, *i.e. in Case (3), firm 1 invests at time* τ_1^{L} *and firm 2 invests at time* τ_2^{F}. *If, on the other hand,* $\mathrm{L}_1^* > \mathrm{F}_2^*$, *i.e. in Case (4), the both firms invest simultaneously at time* τ_2^{F}.

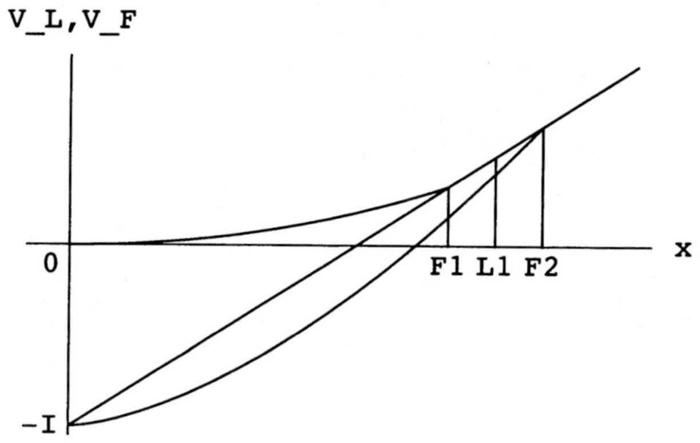

Figure 5. The value functions for firm 1 in Case (3)

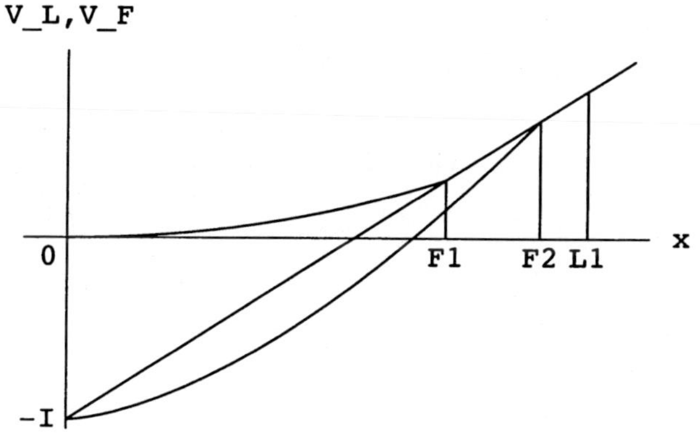

Figure 6. The value functions for firm 1 in Case (4)

5. Conclusions

In this paper, we considered strategic entry decisions of firms in an asymmetric duopoly market when the underlying state variable follows a diffusion with volatility that depends on the current state variable. It is shown that, even in the extended model, there exist three types of equilibria for the strategic substitution case, while there are only two types of equilibria for the strategic complement case.

More specifically, as for the geometric Brownian case, we have: (1) There is a preemption equilibrium for the case of strategic substitution. We find that, when there is a relatively small cost asymmetry among firms, the higher-cost firm may have an incentive to preempt the lower-cost firm. This implies that there is a first-mover advantage in this case. (2) There does not exist any preemption equilibrium for the case of strategic complement. In this situation, there is no incentive to preempt irrespective of the magnitude of relative cost asymmetry among firms.

The impact of market uncertainty is analyzed. Despite the presence of strategic interactions, the increased uncertainty always results in the increase of the firm value. Such a positive relationship is the same as financial options that developed originally by Black and Scholes [3]. Also, we show that the presence of strategic interactions may push a firm with cost advantage to invest earlier, and the optimal threshold for the investment decision increases as the market uncertainty increases.

Acknowledgment

We would like to thank the anonymous referee for helpful comments and discussions.

Appendix

This appendix includes the proofs of our results.

Proof of Lemma 2.1: Under the integrability condition, we have from (2) that

$$(R_r\pi)(x) = \int_0^\infty e^{-rt} E^x[\pi(X_t)]dt, \tag{A.1}$$

by Fubini's theorem. Define

$$C_t(x) = e^{-\mu t} E^x[\pi(X_t)], \quad t \geq 0. \tag{A.2}$$

Then, $C_t(x)$ is thought of as the price of a financial derivative written on the variable $\{X_t\}$ with maturity t, initial value x, the risk-free interest rate μ, and payoff function $\pi(x)$. According to Hobson [9], the price $C_t(x)$ is increasing and convex if so is $\pi(x)$. Since $(R_r\pi)(x) = \int_0^\infty e^{-(r-\mu)t} C_t(x)dt$, the result follows at once.

Proof of Lemma 2.2: Suppose that $\pi(x) = cx + a$ with $c \neq 0$. Then, from (A.1), we have

$$(R_r\pi)(x) = c\int_0^\infty e^{-rt} m_t(x)dt + \frac{a}{r},$$

where $m_t(x) = E^x[X_t]$. On the other hand, from (3), we obtain

$$X_t - x = \mu\int_0^t X_s ds + \int_0^t X_s\sigma(X_s)dz_s, \quad t \geq 0,$$

where $X_0 = x$. Taking the expectation on both sides yields

$$m_t(x) - x = \mu \int_0^t m_s(x)\mathrm{d}s,$$

since the mean of the Ito integral is zero under the given conditions. It follows that $m_t(x) = xe^{\mu t}$, and so

$$(R_r\pi)(x) = cx \int_0^\infty \mathrm{e}^{-(r-\mu)t}\mathrm{d}t + \frac{a}{r} = \frac{c}{r-\mu}x + \frac{a}{r},$$

since $r > \mu$, proving the lemma.

Proof of Lemma 2.3: First, since $h(0) = 0$ and state 0 is absorbing, we obtain $V_T(0) = 0$. Next, let $Y_t = \mathrm{e}^{-\mu t}X_t$. Then, from the SDE (3), we have

$$\text{(A.3)} \qquad \frac{\mathrm{d}Y_t}{Y_t} = \sigma(Y_t, t)\mathrm{d}z_t, \quad t \geq 0,$$

where $\sigma(y, t) = \sigma(\mathrm{e}^{\mu t}y)$, and the value function is rewritten as

$$\text{(A.4)} \qquad V_T(x) = \max_{\tau \in \mathcal{T}} E^x \left[\mathrm{e}^{-r\min\{\tau,T\}} h(\mathrm{e}^{\mu \min\{\tau,T\}} Y_{\min\{\tau,T\}})\right].$$

Note that the function $h(\mathrm{e}^{\mu t}y)$ is increasing and convex in y and the process $\{Y_t\}$ is a martingale.[5] The result follows by the same arguments as in the proof of Theorem 4.1 in Hobson [9].

Proof of Equation (12): From Equation (11), we have

$$\begin{aligned} V_i^{\mathrm{F}}(x) &= E^x\left[\int_t^\infty \mathrm{e}^{-r(s-t)}\pi_{01}(X_s)\mathrm{d}s \middle| \mathcal{F}_t\right] \\ &\quad + \max_{\tau \in \mathcal{T}_t} E^x\left[\int_\tau^\infty \mathrm{e}^{-r(s-t)}\left(\pi_{11}(X_s) - \pi_{01}(X_s)\right)\mathrm{d}s - \mathrm{e}^{-r(\tau-t)}I_i \middle| \mathcal{F}_t\right] \\ &= E^x\left[\int_t^\infty \mathrm{e}^{-r(s-t)}\pi_{01}(X_s)\mathrm{d}s \middle| \mathcal{F}_t\right] \\ &\quad + \max_{\tau \in \mathcal{T}_t} E^x\left[\mathrm{e}^{-r(\tau-t)}\left\{E^x\left[\int_\tau^\infty \mathrm{e}^{-r(s-\tau)}\left(\pi_{11}(X_s) - \pi_{01}(X_s)\right)\mathrm{d}s \middle| \mathcal{F}_\tau\right] - I_i\right\}_+ \middle| \mathcal{F}_t\right], \end{aligned}$$

[5] Kijima [11] showed that the convexity may fail if the process is not martingale. This is the reason why we need to assume that the mean growth rate is constant.

which can be rewritten by the time homogeneity of the diffusion $\{X_t\}$ and its strong Markov property as

$$\begin{aligned} V_i^{\mathrm{F}}(x) &= E^x\left[\int_0^\infty \mathrm{e}^{-rs}\pi_{01}(X_s)\mathrm{d}s\right] \\ &\quad + \max_{\tau\in\mathcal{T}_t} E^x\left[\mathrm{e}^{-r(\tau-t)}\left(E^{X_\tau}\left[\int_0^\infty \mathrm{e}^{-rs}(\pi_{11}(X_s)-\pi_{01}(X_s))\,\mathrm{d}s\right]-I_i\right)\middle|\mathcal{F}_t\right] \\ &= (R_r\pi_{01})(x) + \max_{\tau\in\mathcal{T}_t} E^x\left[\mathrm{e}^{-r(\tau-t)}\{(R_r(\pi_{11}-\pi_{01}))(X_\tau)-I_i\}_+\middle|\mathcal{F}_t\right]. \end{aligned}$$

Proof of Proposition 3.2: For simplicity, we put $t = 0$. Suppose in contrary that $\tau_2^{\mathrm{F}} \le \tau_1^{\mathrm{F}}$. Then, from Equation (12), we have

$$\begin{aligned} V_2^{\mathrm{F}}(x) &= (R_r\pi_{01})(x) + E^x\left[\mathrm{e}^{-r(\tau_2^{\mathrm{F}}-t)}\left((R_r(\pi_{11}-\pi_{01}))(\mathrm{F}_2^*)-I_1\right)\right] \\ &\qquad -(I_2-I_1)E^x\left[\mathrm{e}^{-r(\tau_2^{\mathrm{F}}-t)}\right] \\ &\le V_1^{\mathrm{F}}(x) - (I_2-I_1)E^x\left[\mathrm{e}^{-r(\tau_1^{\mathrm{F}}-t)}\right], \end{aligned}$$

where we put $t = 0$ for notational simplicity. Here, the inequality holds, since τ_1^{F} is optimal for $V_1^{\mathrm{F}}(x)$ and e^{-rt} is decreasing in t. Hence, $\tau_2^{\mathrm{F}} \ge \tau_1^{\mathrm{F}}$. A strict inequality holds, since e^{-rt} is strictly decreasing in t and $V_1^{\mathrm{F}}(x)$ is continuous in x.

References

1. Alvarez, L. H. R. and Stenbacka, R., Adoption of Uncertain Multi-stage Technology Projects: A Real Options Approach, *Journal of Mathematical Economics*, **35**, (2001), 71–97.
2. Bernando, A. E. and Chowdhry, B., Resources, Real Options, and Corporate Strategy, *Journal of Financial Economics*, **63**, (2002), 211–234.
3. Black, F. and Scholes, M., The Pricing of Options and Corporate Liabilities, *Journal of Political Economy*, **81**, (1973), 637–654.
4. Davis, G. A., Estimating Volatility and Dividend Yield When Valuing Real Options to Invest or Abandon, *Quarterly Review of Economics and Finance*, **38**, (1998), 725–754.
5. Dixit, A. K. and Pindyck, R. S., *Investment Under Uncertainty*, (1994), Princeton Unversity Press, Princeton, NJ.
6. Epps, T. W., *Pricing Derivative Securities*, (2002), World Scientific, New York, NY.
7. Fudenberg, D. and Tirole, J., Preemption and Rent Equalization in the Adoption of New Technology, *Review of Economic Studies*, **52**, (1985), 383–401.
8. Geske, R., The Valuation of Compound Options, *Journal of Financial Economics*, **7**, (1979), 63–81.
9. Hobson, D. G., Volatility Misspecification, Option Pricing and Superrepliation via Coupling, *Annals of Applied Probability*, **8**, (1998), 193–205.
10. Huisman, K. J. M., *Technology Investment: A Game Theoretic Real Options Approach*, (2001), Dissertation paper, Tilburg University.

11. Kijima, M., Monotonicity and Convexity of Option Prices Revisited, *Mathematical Finance*, **12**, (2002), 411–426.
12. Kijima, M., *Stochastic Processes with Applications to Finance*, (2002), Chapman & Hall, London.
13. McDonald, R. and Siegel, D. R., The Value of Waiting to Invest, *Quarterly Journal of Economics*, **101**, (1986), 707–727.
14. Pawlina, G. and Kort, P. M., Real Options in an Asymetric Doupoly: Who Benefits from Your Competitive Disadvantage? *Journal of Economics and Management Strategy*, **15**, (2006), 1–35.
15. Shibata, T., The Impacts of Uncertainties in a Real Options Model under Incomplete Information. *European Journal of Operational Research*, **187**, (2008), 1368–1379.
16. Trigeorgis, L., *Real Options: Managerial Flexibility and Strategy in Resource Allocation*, (1996), MIT Press, Cambridge, MA.

Arbitrage Pricing Under Transaction Costs: Continuous Time

Emmanuel Denis

School of Management, Boston University
E-mail: emmanuel.denis6@wanadoo.fr, edenis@bu.edu

We develop an abstract version of Arbitrage Pricing Theory for continuous-time models with transaction costs. Our results includes the financial $\mathcal{Y}$-model of Campi and Schachermayer.
Key words: General arbitrage, Random cones, Continuous trading, Hedging theorem

1. Introduction

Arbitrage in discrete-time models with transaction costs has been initiated in [8] for geometric models defined by random cones and various theorems of asset pricing are provided in a lot of papers as [9], [10], [12].

Arbitrage in continuous-time models with transaction costs is topical. The first work involving continuous-time models is the pioneering paper [6]. The following is the recent Fundamental Theorem of Asset Pricing [4] for continuous processes under small transaction costs which relates a notion of absence of arbitrage to the existence of *consistent price systems*. An other version [5] is proved in a market model with a risky asset. On the other hand, versions of hedging theorems for European options but also American options are produced for continuous-time models in [11], but also in [13] and [1]. In all cases, the formulation of these versions requires the existence of a *consistent price system* which replaces the notion of martingale density in the classical theory without transaction cost. A *consistent price system* $Z \in \mathcal{M}_0^T(G^*)$ is a martingale evolving in the (positive) duals G_t^* of the cones G_t. In the financial context, G_t are solvency cones $\widehat{K}_t$, "hat" means that the assets are measured in physical units (the notation K_t is used for the solvency cones in units of the numéraire; see [7]). That's why, it is of first importance to provide criteria ensuring the existence of such *consistent price systems*.

In the present work, we extend the concept of No Generalized Arbitrage introduced by Cherny in [2] to continuous-time models where the set of attainable incomes at date T is a convex cone $\mathbf{R}_T$ in $L^0(\mathbb{R}^d, \mathcal{F}_T)$, the space of d-dimensional random variables on a probability space $(\Omega, \mathcal{F}_T, P)$ endowed with a filtration $(\mathcal{F}_t)_{t\in[0,T]}$ verifying the usual hypothesis. Recall that Cherny defines the NGA condition only for a set of incomes in $L^0(\mathbb{R}, \mathcal{F}_T)$. The same author applies his theory to a model with transaction costs [3] in the setting of Jouini and Kallal, i.e. with a bid-ask price process but in a one dimensional model where portfolios are expressed in a numeraire. The reasonings in the d-dimensional case and with portfolios expressed in physical units follow the same spirit as the Cherny ones using the notion of Fatou-convergence relating to the admissibility condition required for hedging theorems. We show that the NGA condition is connected to the existence of a *consistent price system* for models whose set of incomes contains the sets $L^0(-G_t, \mathcal{F}_t)$, $t \in [0, T]$. Moreover, in the financial model $\mathcal{Y}$ introduced by Campi and Schachermayer, the *NGA* condition is equivalent to the classical No Free Lunch condition and it is interesting to observe the connection between the "fair prices" defined by Cherny and the "minimal prices" hedging an European Option.

Note that the formulation of hedging theorems requires the condition $\mathcal{D} \neq \emptyset$ where $\mathcal{D} = \mathcal{D}(G)$ is the subset of $\mathcal{M}_0^T(\operatorname{int} G^*)$ formed by martingales Z such that not only $Z_\tau \in \operatorname{int} G_\tau^*$ a.s. for any stopping times τ but also $Z_{\tau-} \in \operatorname{int} G_{\tau-}^*$ a.s. for any predictable times $\tau \in [0, T]$. Finding a criterion ensuring the existence of such a *consistent price system* is an open problem.

2. Generalized Arbitrage in Abstract Setting

Let (Ω, F, P, R_T) be an abstract model where (Ω, F, P) is a probability space with a filtration $F = (\mathcal{F}_t)_{t\in[0,T]}$ and R_T is a cone which is a subset of $L^0(\mathbb{R}^d, \mathcal{F}_T)$. In a financial context, R_T is the set of all terminal values at date T of incomes starting from 0 at date $t = 0$. We define the set $A_T := R_T - L^0(\mathbb{R}^d_+, \mathcal{F}_T)$ of all contingent claims.

We shall give an arbitrage theory inspired by [2]. For this, we introduce some notations.

If $x \in \mathbb{R}^d_+$, we note $x \geq 0$. As in [2], we add to the set R_T the Fatou-convergent sequences of its elements:

$$\overline{R}_T^F := \left\{ \xi = \lim_n \xi^n \text{ a.s} : \exists k \geq 0 \text{ such that } \xi^n \geq -k\mathbf{1},\ \xi^n \in R_T \right\}$$

For $\xi \in \overline{R}_T^F$, we note

$$\Upsilon(\xi) := \mathbf{1} + \xi - \operatorname{essinf} \xi$$

where $\operatorname{essinf} \xi$ is defined componentwise by $(\operatorname{essinf} \xi)^i = \operatorname{essinf} \xi^i$. Observe that for $\xi \in \overline{R}_T^F$, there exists $k \geq 0$ such that $\xi \geq -k\mathbf{1}$. So, $\operatorname{essinf} \xi$ is well defined and $\Upsilon(\xi) \geq \mathbf{1}$ a.s. We note for $a, b \in \mathbb{R}^d$, a/b and $a \times b$ the vectors whose components are respectively a^i/b^i and $a^i b^i$.

We introduce for each $\xi \in \overline{R_T}^F$, $A_{\mathcal{N}}(\xi)$ the set of normalized contingents:

$$A_{\mathcal{N}}(\xi) := \{X/\Upsilon(\xi) : X \in A_T\}, \quad A_{\mathcal{N}}^{\infty}(\xi) := A_{\mathcal{N}}(\xi) \cap L^{\infty}$$

and we note $\overline{A_{\mathcal{N}}^{\infty}}^{w}(\xi)$ the weak closure of $A_{\mathcal{N}}^{\infty}(\xi)$ in $\sigma(L^{\infty}, L^1)$.

Finally, we note $\mathbb{R}^d_{++} = \operatorname{int} \mathbb{R}^d_+$ and we define

$$\begin{aligned}
\mathcal{Z} &:= \{Z \in L^0(\mathbb{R}^d_{++}, \mathcal{F}_T) : E(ZX)^- \geq E(ZX)^+, \ \forall X \in R_T\}, \\
\mathcal{Z}(\xi) &:= \{Z \in L^0(\mathbb{R}^d_{++}, \mathcal{F}_T) : E|ZX| < \infty \text{ and } EZX \leq 0 \\
&\qquad \text{if } X \in R_T \text{ verifies } X \geq -\alpha\xi - \beta\mathbf{1} \text{ where } \alpha, \beta \geq 0\}.
\end{aligned}$$

Definition 2.1. We say that (Ω, F, P, R_T) satisfies the No General Arbitrage property NGA if

$$\overline{A_{\mathcal{N}}^{\infty}}^{w}(\xi) \cap L^0(\mathbb{R}^d_+, \mathcal{F}_T) = \{0\}, \quad \forall \xi \in \overline{R_T}^F.$$

Our main theorem is the following:

Theorem 2.1. *Assume that there exists $\xi_0 \in \overline{R_T}^F$ such that $\mathcal{Z}(\xi_0) = \mathcal{Z}$. Then*

$$NGA \Leftrightarrow \mathcal{Z} \neq \emptyset.$$

To show it, we shall prove later the following results:

Theorem 2.2. *Assume that $\overline{A_{\mathcal{N}}^{\infty}}^{w}(\xi) \cap L^0(\mathbb{R}^d_+) = \{0\}$. Then $\mathcal{Z}(\xi) \neq \emptyset$.*

Theorem 2.3. *Assume that $\mathcal{Z} \neq \emptyset$, then the NGA condition holds.*

We consider a random variable F verifying $F \geq -k_F\mathbf{1}$ for some constant $k_F \geq 0$. In a financial context, F is considered as a contingent claim expressed in physical units. For example, $F(x) = (S_t - K)^+/S_t$ defines the European Call.

Definition 2.2. A real number x is a fair price of F if the extended model $(\Omega, \mathcal{F}, P, R_T + \{h(F - x) : h \in \mathbb{R}\})$ satisfies the NGA condition.

We define I_F the set of all fair prices of F verifying $F \geq -k_F\mathbf{1}$.

Theorem 2.4. *Assume that there exists $\xi_0 \in \overline{R_T}^F$ such that $\mathcal{Z}(\xi_0) = \mathcal{Z}$ and the NGA condition holds. Then:*

$$I_F = \{x \in \mathbb{R}^d : \exists Z \in \mathcal{Z} \text{ such that } EZx = EZF\}.$$

We consider a cone-valued process $G = (G_t)_{t\in[0,T]}$ where we assume that all cones G_t are closed in $\mathbb{R}^d$ and adapted, i.e., their graphs

$$\Delta_t(G) := \left\{(\omega, x) \in \Omega \times \mathbb{R}^d : x \in G_t(w)\right\}$$

are $\mathcal{F}_t$-measurable. We note $G_t^* = \{y \in \mathbb{R}_+^d : yx \geq 0,\ \forall x \in G_t\}$. We define $\mathcal{M}_0^T(G^*)$ the set of all martingales $(Z_t)_{t\in[0,T]}$ such that $Z_t \in G_t^*$, $\forall t \in [0,T]$.

Corollary 2.1. *Assume that $L^\infty(-G_t, \mathcal{F}_t) \subset R_T$, $\forall t \in [0,T]$. Then,*

$$\bigcup_{\xi\in\overline{R}_T^F} \mathcal{Z}(\xi) \cup \mathcal{Z} \neq \emptyset \Rightarrow \mathcal{M}_0^T(G^*) \cap L^0(\mathbb{R}_{++}^d, \mathcal{F}_T) \neq \emptyset.$$

Proof. If $Z_T \in \mathcal{Z}(\xi) \cup \mathcal{Z}$ we define the martingale $Z_t = E(Z_T|\mathcal{F}_t)$. It is easy to prove that for any selector $\zeta_t \in L^\infty(G_t, \mathcal{F}_t)$, the inequality $EZ_t\zeta_t \geq 0$ holds. Then, we can conclude.

Definition 2.3. Assume that $L^\infty(-G_t, \mathcal{F}_t) \subset R_T$, $\forall t \in [0,T]$. We say that (Ω, F, P, R_T) satisfies the Weak No Arbitrage property NA^w if

$$R_T \cap L^0(G_T, \mathcal{F}_T) \subseteq L^0(\partial G_T, \mathcal{F}_T).$$

Recall that G_T dominates $\mathbb{R}_+^d$, means $\mathbb{R}_+^d \setminus \{0\} \subset \operatorname{int} G_T$.

Proposition 2.1. *Assume that G_T dominates $\mathbb{R}_+^d$. Then:*

$$NA^w \Leftrightarrow R_T \cap L^0(\mathbb{R}_+^d, \mathcal{F}_T) = \{0\}.$$

Remark 2.1. $(NGA) \Rightarrow (NA^w)$.

3. Proofs

3.1 Proof of Proposition 2.1

First we assume NA^w and we consider $\xi \in R_T \cap L^0(\mathbb{R}_+^d, \mathcal{F}_T)$. It follows that $\xi \in R_T \cap L^0(G_T, \mathcal{F}_T) \subseteq L^0(\partial G_T, \mathcal{F}_T)$. So, if $\xi \in \mathbb{R}_+^d \setminus \{0\}$ on a non-null set, then $\xi \in \operatorname{int} G_T$ on the latter because the domination of $\mathbb{R}_+^d$ by G_T leads to $\mathbb{R}_+^d \setminus \{0\} \subseteq \operatorname{int} G_T$. This contradicts the fact that $\xi \in \partial G_T$ a.s.

Suppose that $R_T \cap L^0(\mathbb{R}_+^d) = \{0\}$. Let $\xi \in R_T \cap L^0(G_T, \mathcal{F}_T)$ and suppose that $\xi \in \operatorname{int} G_T$ on a non-null set. By a measurable selection argument, we deduce the existence of $X^+ \in L^\infty(\mathbb{R}_+^d, \mathcal{F}_T) \setminus \{0\}$ such that $Z = \xi - X^+$ belongs to $L^0(G_T, \mathcal{F}_T)$. Moreover, there exists n large enough such that we still have $X^+ I_{\|\xi\|\leq n} \in L^\infty(\mathbb{R}_+^d, \mathcal{F}_T) \setminus \{0\}$. Then, $Z^n = \xi - X^+ I_{\|\xi\|\leq n}$ verifies $-Z^n \in L^\infty(-G_T, \mathcal{F}_T)$ and we deduce that $X^+ I_{\|\xi\|\leq n} \in R_T \cap L^0(\mathbb{R}_+^d, \mathcal{F}_T)$ which leads to a contradiction. So, we have

$$R_T \cap L^0(G_T, \mathcal{F}_T) \subseteq L^0(\partial G_T, \mathcal{F}_T).$$

□

3.2 Proof of Theorems 2.2 and 2.3

We need some auxiliary results.
Let recall the following lemma that we can find in [7]:

Lemma 3.1. *Let $\mathcal{G}$ be a family of measurable sets such that any non-null set Γ has the non-null intersection with an element of $\mathcal{G}$. Then, there exists a countable subfamily of sets $\{\Gamma_i\}$ of full measure.*

We can deduce the following theorem:

Theorem 3.1. *Let C be a convex cone in L^∞ closed in $\sigma(L^\infty, L^1)$ containing $L^\infty(\mathbb{R}^d_-)$ and such that $C \cap L^\infty(\mathbb{R}^d_+) = \{0\}$. Then, there exists $\rho \in L^0(\mathbb{R}^d_{++})$ verifying $E\rho X \leq 0$ for all $X \in C$ and $E|\rho| < \infty$.*

Proof. We deduce from the Hahn-Banach theorem that for any element $x \in L^\infty(\mathbb{R}^d_+) \setminus \{0\}$, there exists $Z_x \in L^1$ such that we have $EZ_x\xi < EZ_x x$, for all $\xi \in C$. We note e^i the vector whose only the ith component is non-null and equal to unit. Taking $\xi = \alpha e^i Z^i_x 1_{-M\leq Z^i_x<0} \in C$, for all $\alpha, M \geq 0$ and $i = 1, \cdots, d$, we deduce that $Z_x \geq 0$. Moreover, $Z_x \neq 0$. So, we can assume that $EZ_x \leq \mathbf{1}$ and $EZ_x x > 0$. Let define

$$\mathcal{G}^k := \left\{\{Z^k_x \neq 0\},\ x \in L^\infty(\mathbb{R}^d_+) \setminus \{0\}\right\}.$$

Then, for all Γ such that $P(\Gamma) \neq 0$, we have $P(\Gamma \cap \{Z^k_x \neq 0\}) \neq 0$ where $x = e^k 1_\Gamma$. Indeed, $EZ_x.x > 0$. We deduce from the previous lemma a countable family $Z_{x_{k,i}}$ such that

$$P\left(\cup_i \left\{Z^k_{x_{k,i}} \neq 0\right\}\right) = 1.$$

Defining $\rho = \sum_{k,i} 2^{-k-i} Z_{x_{k,i}}$, it is obvious that $E|\rho| < \infty$ and we can easily verify that for all k, $\rho^k > 0$ on the set

$$N^c := \cap_{k=1}^d \cup_i \left\{Z^k_{x_{k,i}} \neq 0\right\}$$

of full measure. So $\rho \in \mathbb{R}^d_{++}$ a.s. and from above we have $E\rho\xi \leq 0$ for all $\xi \in C$. □

Lemma 3.2. *For all $\xi \in \overline{R}^F_T$, we have $\mathcal{Z} \subseteq \mathcal{Z}(\xi)$.*

Proof. Let consider $\xi \in \overline{R}^F_T$ and $Z \in \mathcal{Z}$. Recall that $\xi = \lim_n \xi_n$ where $\xi_n \in R_T$ verifies $\xi_n \geq -k\mathbf{1}$, $k \geq 0$.

Since $Z \in \mathcal{Z}$, we have the inequality $E(Z\xi_n)^- \geq E(Z\xi_n)^+$ whereas $Z\xi_n \geq -kZ\mathbf{1}$ implies that $E(Z\xi_n)^- \leq kE|Z\mathbf{1}| < \infty$. It follows that $E|Z\xi_n| \leq 2kE|Z\mathbf{1}|$ and in virtue of the Fatou lemma, we deduce that $E|Z\xi| < \infty$.

From now on, we consider $X \in R_T$ verifying $X \geq -\alpha\xi - \beta\mathbf{1}$ with $\alpha, \beta \geq 0$. Then, $ZX \geq -\alpha Z\xi - \beta Z\mathbf{1}$ and it follows that $E(ZX)^- < \infty$.

So, $Z \in \mathcal{Z}$ implies that $E(ZX)^+ \leq E(ZX)^- < \infty$ and $EZX \leq 0$. We can conclude that $Z \in \mathcal{Z}(\xi)$. □

Lemma 3.3. *Let consider $\xi_0 \in \overline{R_T}^F$ such that there exists $\rho \in L^0(\mathbb{R}^d_{++})$ verifying $E|\rho| < \infty$ and $E\rho X \leq 0$ for all $X \in \overline{A^\infty_\mathcal{N}}^w(\xi_0)$. Assume that $X \in R_T$ verifies $X/\Upsilon(\xi_0) \geq -a\mathbf{1}$ where $a \geq 0$. Then we have:*

$$E\left|\rho\frac{X}{\Upsilon(\xi_0)}\right| < \infty, \quad E\rho\frac{X}{\Upsilon(\xi_0)} \leq 0.$$

Proof. For all $c \in \mathbb{R}_+$, we define $(X - c\mathbf{1})^+$ and $X \wedge c\mathbf{1}$ the random variables whose components are respectively $(X^i - c)^+$ and $X^i \wedge c$. Observe that $X \wedge c\mathbf{1} = X - (X - c\mathbf{1})^+ \in A_T$. It follows that $\zeta_c = X \wedge c\mathbf{1}/\Upsilon(\xi_0) \in A_\mathcal{N}(\xi_0)$. Moreover, $-a\mathbf{1} \leq \zeta_c \leq c\mathbf{1}$. So $\zeta_c \in \overline{A^\infty_\mathcal{N}}^w(\xi_0)$ and $E\rho\zeta_c \leq 0$. But we have

$$\rho\zeta_c = \sum_i \rho^i\zeta_c^i \geq -a\sum_i \rho^i$$

where $E\rho^i < \infty$. So, we can apply the Fatou lemma as $c \to \infty$ to conclude about the lemma. □

Corollary 3.1. *Assume that $\overline{A^\infty_\mathcal{N}}(\xi_0) \cap L^0(\mathbb{R}^d_+) = \{0\}$. Then $\mathcal{Z}(\xi_0) \neq \emptyset$.*

Proof. We claim that $L^\infty(-\mathbb{R}^d_+, \mathcal{F}_T) \subseteq \overline{A^\infty_\mathcal{N}}^w(\xi_0)$. If $X \in L^\infty(-\mathbb{R}^d_+, \mathcal{F}_T)$, it suffices to observe that $\xi = \Upsilon(\xi_0) \times X \in L^0(-\mathbb{R}^d_+, \mathcal{F}_T)$, which implies that $X \in A_T$ and finally $X \in \overline{A^\infty_\mathcal{N}}^w(\xi_0)$.

From Theorem 3.1, we deduce the existence of $\rho \in L^0(\mathbb{R}^d_{++})$ such that $E|\rho| < \infty$ and $E\rho X \leq 0$ for all $X \in \overline{A^\infty_\mathcal{N}}^w(\xi_0)$.

Let's show that $\rho \in \mathcal{Z}(\xi_0)$. If $X \in R_T$ verifies $X \geq -\alpha\xi_0 - \beta\mathbf{1}$ where $\alpha, \beta \geq 0$, we have:

$$\frac{X^i}{\Upsilon(\xi_0)^i} \geq -\alpha + \frac{\alpha(1 - \operatorname{essinf} \xi_0^i) - \beta}{\Upsilon(\xi_0)^i} \geq -\alpha$$

if $\alpha(1 - \operatorname{essinf} \xi_0^i) - \beta \geq 0$ and otherwise

$$\frac{\alpha(1 - \operatorname{essinf} \xi_0^i) - \beta}{\Upsilon(\xi_0)^i} \geq \alpha(1 - \operatorname{essinf} \xi_0^i) - \beta.$$

So, there exists $a \geq 0$ such that $X/\Upsilon(\xi_0) \geq -a\mathbf{1}$. From Lemma 3.3, we deduce that $E|\rho X| < \infty$ and $E\rho X \leq 0$. Thus, $\rho \in \mathcal{Z}(\xi_0)$. □

Corollary 3.2. *Assume that $\mathcal{Z} \neq \emptyset$. Then, the NGA condition holds.*

Proof. Suppose $\mathcal{Z} \neq \emptyset$ and consider $Z \in \mathcal{Z} \subseteq \mathcal{Z}(\xi)$ for all $\xi \in \overline{R}_T^F$.
If $Y \in \overline{A_N^{\infty}}^{w}(\xi) \cap L^0(\mathbb{R}_+^d)$, we have $Y = \lim Y_n$ in $\sigma(L^\infty, L^1)$ where

$$Y_n = \frac{X_n - \varepsilon_n}{\Upsilon(\xi)}, \quad X_n \in R_T, \quad \varepsilon_n \geq 0.$$

We deduce that $X_n \geq -\|Y_n\|_\infty \Upsilon(\xi)$. Since $Z \in \mathcal{Z}(\xi)$ we deduce that $E|ZX_n| < \infty$ and $EZX_n \leq 0$. In the proof of Lemma 3.2, we have shown that $E|Z\xi| < \infty$. Moreover, the components of $Z \times \Upsilon(\xi)$ verify the inequalities $0 \leq Z^i \Upsilon(\xi)^i \leq Z\Upsilon(\xi)$. It follows that $Z \times \Upsilon(\xi) \in L^1$ and by hypothesis $EZ \times \Upsilon(\xi) Y_n \to EZ \times \Upsilon(\xi) Y$. We deduce that $EZ \times \Upsilon(\xi) Y \leq 0$ and finally $Y = 0$ which proves the NGA condition. □

3.3 Proof of Theorem 2.4

If we define $\xi_0' := \xi_0 + F - x$, we have obviously $\xi_0' \in \overline{R}_T^{F\prime}$ where $\overline{R}_T^{F\prime}$ is the set analogously defined as $\overline{R}_T^F$ but relatively to $R_T' := R_T + \{h(F - x) : h \in \mathbb{R}\}$. In a similar way, we note $A_T' := R_T' - L^0(\mathbb{R}_+^d, \mathcal{F}_T)$.

Since $x \in I_F$ implies the NGA condition for the extended model, we deduce, from Theorem 3.1, the existence of a random variable $\rho \in L^0(\mathbb{R}_{++}^d)$ verifying $E|\rho| < \infty$ and $E\rho X \leq 0$ for all $X \in \overline{A_N^{\infty\prime}}^{w}(\xi_0')$. From Lemma 3.3, we deduce that if $X \in A_T'$ verifies $X/\Upsilon(\xi_0') \geq -a\mathbf{1}$ where $a \geq 0$, then we have $E|\rho X/\Upsilon(\xi_0')| < \infty$ and $E\rho X/\Upsilon(\xi_0') \leq 0$.

We define $Z := \rho/\Gamma(\xi_0')$. From the inclusion $R_T \subset R_T'$, it is easy to see that $Z \in \mathcal{Z}(\xi_0)$ in virtue of the reasoning proving Corollary 3.1. Moreover, since we assume that F is bounded from below, we deduce from what precedes that $F - x \in A_T'$ implies that $E|Z(F - x)| < \infty$ and $EZ(F - x) \leq 0$. On the other hand, if $b \in \mathbb{R}_+$, from $(x - F)I_{x-F \geq b\mathbf{1}} = x - F - (x - F)I_{x-F \geq b\mathbf{1}}$ where $x - F \in R_T'$, we deduce that $(x - F)I_{x-F \geq b\mathbf{1}} \in A_T'$ and we obtain that $EZ(x - F)I_{x-F \geq b\mathbf{1}} \leq 0$. It follows that $EZ(x - F) \leq EZ(x - F)I_{x-F \geq b\mathbf{1}}$ where we recall that $E|Z(F - x)| < \infty$. Using the Lebesgue Theorem, we deduce that $EZ(x - F) \leq 0$ as $b \to \infty$ and finally $EZ(x - F) = 0$.

Reciprocally, suppose that $EZx = EZF$ for some $Z \in \mathcal{Z}$. First, we consider $Y = X + h(F - x) \in R_T'$ with $Y \geq -\alpha\mathbf{1}$ for some constant $\alpha \geq 0$. We deduce easily that $E(ZX)^- < \infty$ and finally from $Z \in \mathcal{Z}$, we have $EZX \leq 0$. So, $E|ZY| < \infty$ and $EZY \leq 0$. It follows that for all $\xi' \in \overline{R}_T^{F\prime}$, we have $E|Z\xi'| < \infty$ and $EZ\xi' \leq 0$ because of the Fatou lemma.

From now on, we define $\overline{Z} := \Upsilon(\xi') \times Z$. From what precedes, we deduce that $E|\overline{Z}| < \infty$. Indeed, for any $i = 1, \cdots, d$, $E|Z^i \xi'^i| < \infty$ and recall that $0 \leq Z^i \Upsilon(\xi')^i = Z^i(1 + \xi'^i - \operatorname{essinf} \xi'^i)$. We deduce that if an element $Y = X + h(F - x) \in R_T'$ verifies the inequality $Y/\Upsilon(\xi') \geq -\alpha\mathbf{1}$, then we have

$$E|\overline{Z}\frac{Y}{\Upsilon(\xi')}| < \infty, \quad E\overline{Z}\frac{Y}{\Upsilon(\xi')} \leq 0.$$

Indeed, from $Y/\Upsilon(\xi') \geq -\alpha\mathbf{1}$, we first deduce two constants $\alpha, \beta \geq 0$ such that $Y \geq -\alpha\xi' - \beta\mathbf{1}$ and from $E|Z\xi'| < \infty$, we obtain the inequality $E(ZX)^- < \infty$. Since $Z \in \mathcal{Z}$, it follows that $E|ZX| < \infty$ and $EZX \leq 0$. Finally,

$$E\overline{Z}\frac{Y - V}{\Upsilon(\xi')} \leq 0$$

provided that $(Y - V)/\Upsilon(\xi') \geq -\alpha\mathbf{1}$ and $V \geq 0$. It follows that $E\overline{Z}X \leq 0$ for all $X \in A_{\mathcal{N}}^{\infty'}(\xi')$ and finally for all $X \in \overline{A_{\mathcal{N}}^{\infty'}}^{w}(\xi')$. But, if we consider $X \geq 0$ with $X \in \overline{A_{\mathcal{N}}^{\infty'}}^{w}(\xi')$, we have obviously $\overline{Z}X \geq 0$. So, $\overline{Z}X = 0$ and $X = 0$ since $\overline{Z} \in \mathbb{R}^d_{++}$ a.s. Thus, the NGA condition is verified and $x \in I_F$. □

4. Application to Finance

4.1 The Y-Model

In this section, we shall consider the Campi–Schachermayer model Y defined as follows (see [7]).

We are given on the interval $[0, T]$ two set-valued processes $G = (G_t)$ and $G^* = (G_t^*)$ where $G_t = \text{cone}\,\{\xi_t^k : k \in \mathbb{N}\}$ and $G_t^* = \text{cone}\,\{\zeta_t^k : k \in \mathbb{N}\}$. It is assumed that the generating processes are càdlàg, adapted, and for each ω only a finite number of $\xi_t(\omega)$, $\xi_{t-}^k(\omega)$, $\zeta_t^k(\omega)$ and $\zeta_{t-}^k(\omega)$ are different from zero, i.e. all cones are polyhedral. We put $G_{t-}^* = \text{cone}\,\{\zeta_{t-}^k : k \in \mathbb{N}\}$.

Let Y be a d-dimensional predictable process of bounded variation starting from zero and having trajectories with left and right limits (French abbreviation: *làdlàg*). Put $\Delta Y := Y - Y_-$, as usual, and $\Delta^+ Y := Y_+ - Y$ where $Y_+ = (Y_{t+})$. Define the right-continuous processes

$$Y_t^d := \sum_{s \leq t} \Delta Y_s, \qquad Y_t^{d,+} := \sum_{s \leq t} \Delta^+ Y_s$$

(the first is predictable while the second is only adapted) and, at last, the continuous one:

$$Y^c := Y - Y^d - Y_-^{d,+}.$$

Let $\mathcal{Y}$ be the set of such process Y satisfying the following conditions:

1) $\dot{Y}^c \in -G \; dP \, d\|Y^c\|$-a.e.;

2) $\Delta^+ Y_\tau \in -G_\tau$ a.s. whatever is a stopping time $\tau \leq T$;

3) $\Delta Y_\sigma \in -G_{\sigma-}$ a.s. whatever is a predictable stopping time $\sigma \leq T$.

Let $\mathcal{Y}^x := x + \mathcal{Y}$, $x \in \mathbb{R}^d$. We denote by $\mathcal{Y}_b^x$ the subset of $\mathcal{Y}^x$ formed by the processes Y such that $Y_t + \kappa_Y \mathbf{1} \in L^0(G_t, \mathcal{F}_t)$, $t \leq T$, for some $\kappa_Y \in \mathbb{R}$. In the financial context (where $G = \widehat{K}$) the elements of $\mathcal{Y}_b^x$ are the *admissible* portfolio processes.

We associate with Y the following right-continuous adapted process of bounded variation:

$$\bar{Y} := Y^c + Y^d + Y^{d,+},$$

i.e. $\bar{Y} = Y + \Delta^+ Y = Y_+$. Since the generators are right-continuous, the process $\bar{Y}$ inherits the boundedness from below of Y (by the same constant process $\kappa_Y \mathbf{1}$).

We recall the following lemma [13]

Lemma 4.1. *If $Z \in \mathcal{M}_0^T(G^*)$ and $Y \in \mathcal{Y}_b^x$, then the process ZY is a supermartingale and*

$$E(-Z\dot{\bar{Y}}) \cdot \|\bar{Y}\|_T \leq Z_0 x - EZ_T \bar{Y}_T. \tag{4.1}$$

4.2 Generalized Arbitrage for the Y-Model

From Lemma 4.1, it follows that

$$\mathcal{Z} = \mathcal{Z}(\xi) = \left\{ Z_T : Z \in \mathcal{M}_T^0(G^* \cap \mathbb{R}_{++}^d) \right\}$$

for any $\xi \in \overline{R_T}^F$, in the case where $R_T = \mathcal{Y}_b^0(T)$. If we only assume that $R_T = \widetilde{\mathcal{Y}}^0(T)$, where

$$\widetilde{\mathcal{Y}}^0(T) := \left\{ Y \in \mathcal{Y}^0(T) \text{ such that } Z \in \mathcal{M}_T^0(G^*) \Rightarrow \exists M \in \mathcal{M} : ZY \geq M \right\}$$

and $\mathcal{M}$ is the set of all real-valued martingales, then it is always true:

Lemma 4.2. *Assume that $R_T = \widetilde{\mathcal{Y}}^0(T)$ in the Campi–Schachermayer model Y. Then, for any $\xi \in \overline{R_T}^F$,*

$$\mathcal{Z} = \mathcal{Z}(\xi) = \left\{ Z_T : Z \in \mathcal{M}_T^0(G^* \cap \mathbb{R}_{++}^d) \right\}.$$

Proof. Recall that $\mathcal{Z} \subseteq \mathcal{Z}(\xi)$. From now on, we fix $Z_T \in \mathcal{Z}(\xi)$ and we define the martingale $Z_t = E(Z_T | \mathcal{F}_t)$ which belongs to $\mathcal{M}_T^0(G^* \cap \mathbb{R}_{++}^d)$. The proof of Lemma 4.1 is based on the following product formula:

$$Z_t Y_t = Z_0 x + Y \cdot Z_t + Z\dot{Y}^c \cdot \|Y^c\|_t + \sum_{s \leq t} Z_{s-} \Delta Y_s + \sum_{s < t} Z_s \Delta^+ Y_s$$

where $Y \in R_T$. We deduce that $N_t = Y \cdot Z_t$ is a local martingale bounded from below by a martingale M because of the hypothesis $Y \in \widetilde{\mathcal{Y}}^0(T)$. It follows that N is a supermartingale and Lemma 4.1 holds for $Y \in \widetilde{\mathcal{Y}}^0(T)$, i.e. $|EZ_T Y_T| < \infty$ and $EZ_T Y_T \leq 0$. □

We deduce easily the following

Lemma 4.3. *Assume that $R_T = \mathcal{Y}_b^0(T)$ or $R_T = \widetilde{\mathcal{Y}}^0(T)$ in the Campi–Schachermayer model Y. Then, the following conditions are equivalent:*

a) NGA

b) $\mathcal{M}_T^0(G^* \cap \mathbb{R}_{++}^d) \neq \emptyset$

c) $\overline{A_N^\infty}^w(0) \cap L^0(\mathbb{R}_+^d) = \{0\}$

We recall the dual description of the initial endowments of portfolios hedging a contingent claim F_T defining a European Call and verifying the admissible condition $F_T + k\mathbf{1} \geq 0$ where $k \geq 0$. For this, we define the convex set

$$\Gamma_F := \left\{x \in \mathbb{R}^d : F \in \mathcal{Y}_b^x(T)\right\}$$

and the closed convex set

$$D_F := \left\{x \in \mathbb{R}^d : Z_0 x \geq E Z_T F \quad \forall Z \in \mathcal{M}_0^T(G^*)\right\}.$$

We denote by $\mathcal{D} := \mathcal{D}(G)$ the subset of $\mathcal{M}_0^T(\operatorname{int} G^*)$ formed by martingales Z such that not only $Z_\tau \in \operatorname{int} G_\tau^*$ a.s. for any stopping times τ but also $Z_{\tau-} \in \operatorname{int} G_{\tau-}^*$ a.s. for any predictable times $\tau \in [0, T]$.

From now on, it is assumed that the cones G_t dominate $\mathbb{R}_+^d$, i.e. we have $\mathbb{R}_+^d \backslash \{0\} \subset \operatorname{int} G_t$ but also the cones G_t are proper, i.e. we have the *efficient friction* condition $G_t^0 := G_t \cap (-G_t) = \{0\}$.

Note that $\mathcal{D} \neq \emptyset$ implies the (NGA) condition. We recall the following version of hedging theorem:

Theorem 4.1. *Assume $\mathcal{D} \neq \emptyset$, G dominates $\mathbb{R}_+^d$ and $G^0 = \{0\}$. Then, we have $\Gamma_F = D_F$.*

The following lemma introduces the notion of minimal prices for a given European option whose contingent claim is F verifying the admissibility condition $F \geq -k_F \mathbf{1}$ where $k_F \in \mathbb{R}^+$. The very natural idea is to define the "smallest" prices $p_F \in \Gamma_F$ of F according to the partial ordering induced by $\mathbb{R}_+^d$.

Lemma 4.4. *Assume $\mathcal{D} \neq \emptyset$, G dominates $\mathbb{R}_+^d$ and $G^0 = \{0\}$. If $\Gamma_F \neq \emptyset$, we can define minimal prices $p_F \in \Gamma_F$ according to the partial ordering generated by $\mathbb{R}_+^d$.*

Proof. By the Zorn lemma, it is easy to show that the set $\Gamma_F \neq \emptyset$ has, at least, a minimal element p_F. Indeed, it suffices to prove that Γ_F is inductive. For this, we consider a subset E totally ordered and we prove that it has a minorant belonging to Γ_F. The case where E is finite is obvious. In the other case, we define $\overline{x}$ componentwise by $\overline{x}_i = \inf_{x \in E} x_i$. It follows that there are sequences $n^i(p)$, $i = 1, \cdots, d$ such that $\overline{x}_i = \lim \searrow x_i^{n^i(p)}$ where $x_i^{n^i(p)} \in E$. Since E is totally ordered, $x^{n^1(p)} \leq x^{n^2(p)}$ or $x^{n^2(p)} \leq x^{n^1(p)}$. Assume for example that $x^{n^1(p')} \leq x^{n^2(p')}$ for a subsequence p'. Then, we have also $\overline{x}_2 = \lim \searrow x_2^{n^1(p')}$ and we can replace n^1 and n^2 by a common subsequence. Inductively, we deduce a common sequence $m(n)$ such that $\overline{x} = \lim_n x^{m(n)}$ where $x^{m(n)} \in E$. Since Γ_F is closed in virtue of Theorem 4.1, it follows that $\overline{x}$ is a minorant for E. □

From the definition, $p_F \in \Gamma_F$ and we have the following characterization:

Theorem 4.2. *Assume $\mathcal{D} \neq \emptyset$, G dominates $\mathbb{R}^d_+$ and $G^0 = \{0\}$. Suppose that $\Gamma_F \neq \emptyset$, then the following two conditions are equivalent:*

a) $p_F \in \Gamma_F$ is a minimal price.

b) There exists a sequence $Z^n \in \mathcal{M}_0^T(G^*\backslash\{0\})$ verifying $Z_0^n \to Z_0$ where $|Z_0| = 1$ and $EZ_T^n F \to Z_0 p_F$.

Proof. Assume that $p_F \in \Gamma_F$ is a minimal price. Thus, $p_F - \frac{1}{n}\mathbf{1} \notin \Gamma_F$ implies that there exists $Z^n \in \mathcal{M}_T^0(G^*\backslash\{0\})$ verifying $|Z_0^n| = 1$ and

$$Z_0^n(p_F - \frac{1}{n}\mathbf{1}) < EZ_T^n F \leq Z_0^n p_F.$$

Since $Z_0^n \to Z_0$ for a subsequence, we deduce easily that $EZ_T^n F \to Z_0 p_F$.

Reciprocally, if x verifies $x \leq p_F$ and $x \in \Gamma_F$, then

$$Z_0^n p_F \geq Z_0^n x \geq EZ_T^n F$$

and as $n \to \infty$, we obtain that $Z_0(p_F - x) = 0$. Moreover, $p_F - x \geq 0$ and

$$Z_0 \in G_0^*\backslash\{0\} \subseteq \mathbb{R}^d_{++}$$

since G dominates $\mathbb{R}^d_+$. Thus $x = p_F$ and p_F is a minimal price. □

We note Γ_F^m the set of all minimal prices of Γ_F.

Corollary 4.1. *Assume $\mathcal{D} \neq \emptyset$, G dominates $\mathbb{R}^d_+$ and $G^0 = \{0\}$. Then,*

$$\partial\Gamma_F = \Gamma_F^m.$$

Proof. Let consider $x \in \partial\Gamma_F$. Then, $x = \lim_n x^n$ where $x^n \notin \Gamma_F$. It follows that there exists $Z^n \in \mathcal{M}_0^T(G^*)$ with $|Z_0^n| = 1$ such that $Z_0^n x^n < EZ_T^n F$. We can assume that $Z_0^n \to Z_0$ with $|Z_0| = 1$. From $Z_0^n x \geq EZ_T^n F \geq Z_0^n x^n$, we deduce that $EZ_T^n F \to Z_0 x$ and we conclude that x is a minimal price in virtue of Theorem 4.2. Reciprocally, assume that $x \in \Gamma_F^m$ is such that $x \notin \partial\Gamma_F$. It follows that $x \in \operatorname{int}\Gamma_F$ and we can deduce that for n large enough $x - n^{-1}\mathbf{1} \in \Gamma_F$ which contradicts the fact that x is a minimal price. □

Then, we deduce:

Lemma 4.5. *Assume that $\mathcal{D} \neq \emptyset$ and G dominates $\mathbb{R}^d_+$. Then,*

$$\overline{\Gamma_F \cap I_F} \subset \Gamma_F^m.$$

Proof. It is obvious since $\Gamma_F \cap I_F \subset \Gamma_F^m$ and Γ_F^m is closed. □

Remark 4.1. It is clear that if I_F is closed, then $\Gamma_F \cap I_F = \Gamma_F^m$.

The following theorem shows that the set I_F is connected to the minimal prices Γ_F^m of Γ_F. Note that

$$I_F = \{x \in \mathbb{R}^d : \exists Z \in \mathcal{M}_T^0(G^*\setminus\{0\}) \text{ such that } Z_0 x = EZ_T F\}.$$

Theorem 4.3. *Assume that $\mathcal{D} \neq \emptyset$ and G dominates $\mathbb{R}_+^d$. Suppose that $e \in \mathbb{R}^d$ is such that $e = \lim_{n\to\infty} \nearrow x_n$ where $x_n \in I_F$. Then, $e \notin I_F$ implies that e is a minimal price of Γ_F.*

Proof. We consider the extended model defined in Definition 2.2 for e. Note that we have obviously $\mathcal{Z}' = \mathcal{Z}'(0)$ because F is bounded. So, it suffices to prove that

$$\overline{A_{\mathcal{N}}^{\infty'}}^{w}(0) \cap L^0(\mathbb{R}_+^d) = \{0\}$$

in order to have $\mathcal{Z}' \neq \emptyset$ and prove the NGA condition.

Let be $\xi \in \overline{A_{\mathcal{N}}^{\infty'}}^{w}(0) \cap L^0(\mathbb{R}_+^d)$. Recall that we have the weak convergence $\xi = \lim(r_T^m - \varepsilon^m)$ where $\varepsilon^m \geq 0$. Moreover,

$$r_T^m = X_T^m + h_m(F - e) \in R_T'^e$$

where $X_T^m \in \mathcal{Y}_b^0(T)$ and $R_T'^e$ is the set of all the incomes corresponding to the extended model for e. Considering

$$r_T^{m,n} = X_T^m + h_m(F - x_n) \in R_T'^{x_n},$$

we can assume that $\xi = \lim(r_T^m - \varepsilon^m)$ where

$$r_T^m = X_T^m + h_m(F - x_m) \in R_T'^{x_m}$$

and $x_m \nearrow e$. In the case where $h_m \geq 0$, we deduce (at least for a subsequence) that for $m \geq m_0$,

$$r_T^m \leq X_T^m + h_m(F - x_{m_0}).$$

Since $\widetilde{Z}_0 x_{m_0} = E\widetilde{Z}_T F$ for some $\widetilde{Z} \in \mathcal{M}_T^0(G^*\setminus\{0\})$, we deduce the inequalities $Er_T^m\widetilde{Z}_T \leq 0$ and $E\widetilde{Z}_T\xi \leq 0$ as $n \to \infty$. Thus $\xi = 0$.

Otherwise, assume that $h_m \leq 0$. If $e \notin \Gamma_F$, there exists $Z \in \mathcal{M}_T^0(G^*\setminus\{0\})$ such that $Z_0 e \leq EZ_T F$. We deduce that

$$h_m EZ_T(F - x_m) \leq h_m EZ_T(F - e) \leq 0.$$

So, we have also $\xi = 0$ and $e \in I_F$.

Finally, we can suppose that $e \in \Gamma_F$. Since we have for some $Z^n \in \mathcal{M}_T^0(G^*\setminus\{0\})$,

$$EZ_T^n F = Z_0^n x_n \to Z_0 e$$

with $Z_0^n \to Z_0$ and $|Z_0^n| = 1$, it suffices to apply Theorem 4.2 to conclude that e is a minimal price. □

Remark 4.2. Observe that $x \in \Gamma_F$ and $Z \in \mathcal{M}_T^0(G^*\setminus\{0\})$ implies that the following vector

$$I(x) := x - \frac{Z_0 x - EZ_T F}{Z_0\mathbf{1}}\mathbf{1}$$

belongs to I_F since we have $Z_0 I(x) = EZ_T F$. Recall that for any minimal price x, there exists a sequence $Z^n \in \mathcal{M}_0^T(G^*\setminus\{0\})$ verifying $Z_0^n \to Z_0$ where $|Z_0| = 1$ and $EZ_T^n F \to Z_0 p_F$ in virtue of Theorem 4.2. It follows that $x^n = x - \alpha^n \mathbf{1} \to x$ where $\alpha^n = (Z_0^n x - EZ_T^n F)/Z_0^n \mathbf{1} \to 0$. Moreover, since $x \in \Gamma_F$, $\alpha^n \geq 0$ and finally, we can assume that α^n is decreasing and $x = \lim \nearrow x^n$ where $x^n \in I_F$.

From the last remark, it is easy to establish the following corollary.

Corollary 4.2. *Assume that $\mathcal{D} \neq \emptyset$ and G dominates $\mathbb{R}_+^d$. Then,*

$$\Gamma_F^m = \{x \in \Gamma_F : x = \lim \nearrow x^n,\ x^n \in I_F\}.$$

4.3 Examples

Example 1. We consider a d-dimensional $\mathcal{Y}$-model with the price process $S = (S^1, \cdots, S^d)$ where S^i, $i = 1, \cdots, d$ are continuous and strictly positive martingales. We assume that the transaction costs $\lambda_t^{i,j} > 0$ are continuous. It is clear that $S \in \mathcal{D}$. We consider the contingent claim $F = \mathbf{1}$. We have obviously $EZ_T F = Z_0\mathbf{1}$ for any $Z \in \mathcal{M}_0^T(G^*)$. Then, $\mathbf{1} \in I_F \cap \Gamma_F$ and $\mathbf{1}$ is a minimal price.

Example 2. We consider a two-dimensional $\mathcal{Y}$-model with the price process $S_t = (S_t^1, S_t^2)_{t\in[0,1]}$ where $S^1 = 1$ and $S_t^2 = I_{[0,1[} + \xi I\{1\}(t)$ is defined with a strictly positive random variable ξ verifying $E\xi = E\xi^2$. Note that a random variable ξ whose density is $f(x) = 2/3 I_{]0,3/2]}(x)$ is appropriated. The filtration is $\mathcal{F}_t = \mathcal{F}_t^S$, $t \in [0, 1]$. In a financial context, G is the solvency cone

$$\widehat{K}_t := cone\{\pi_t^{i,j} e^i - e^j, \quad i, j = 1, 2\}$$

where

$$\pi_t^{1,2} := (1 + \lambda) S_t^2, \quad \pi_t^{2,1} := \frac{1 + \lambda}{S_t^2}$$

since we suppose that the transaction costs coefficients are constant and equal to $\lambda > 0$. It is easily seen that for any $t \in [0, 1[$,

$$G_t = \text{cone}\left\{-e^1 + (1 + \lambda)e^2, (1 + \lambda)e^1 - e^2\right\}$$

whereas

$$G_1 = \text{cone}\left\{-e^1 + (1 + \lambda)\xi e^2, \frac{1 + \lambda}{\xi} e^1 - e^2\right\}.$$

We deduce that for any $t \in [0, 1[$,

$$G_t^* = \text{cone}\left\{(1+\lambda)e^1 + e^2, e^1 + (1+\lambda)e^2\right\}$$

and

$$G_1^* = \text{cone}\left\{(1+\lambda)\xi e^1 + e^2, \xi e^1 + (1+\lambda)e^2\right\}.$$

Observe that $Z_t = E(Z_T|\mathcal{F}_t)$ with $Z_T = (\xi, 1)$ is a martingale in $\mathcal{D}$. Moreover, a martingale whose terminal value is $Z_T = (f(\xi), g(\xi))$, $f, g > 0$, belongs to $\mathcal{M}_0^T(G^*)$ if and only if the two following property holds

$$\frac{f(\xi)}{1+\lambda} \leq g(\xi)\xi \leq (1+\lambda)f(\xi), \tag{4.2}$$

$$\frac{Ef(\xi)}{1+\lambda} \leq Eg(\xi) \leq (1+\lambda)Ef(\xi). \tag{4.3}$$

We remark that for any $g \in G_0^*$, there exists a martingale $Z \in \mathcal{M}_0^T(G^*)$ such that $Z_0 = g$. The case $g = 0$ is obvious. We can assume that the first component is $g^{(1)} = 1$ since G_0^* is a cone. It follows that $g^{(2)} \in [(1+\lambda)^{-1}, 1+\lambda]$. We define $Z_T := (f(\xi), g(\xi))$ with

$$f(\xi) := \frac{1}{E\xi}\xi^2, \quad g(\xi) := \frac{g^{(2)}}{E\xi}\xi.$$

It is easy to verify that the martingale $Z_t = E(Z_T|\mathcal{F}_t)$ verifies Conditions (4.2) and (4.3) which implies that $Z \in \mathcal{M}_0^T(G^*)$ with $Z_0 = g$.

We consider the contingent claim $F := (0, 1)$ defining a European option. We recall that

$$\Gamma_F := \left\{x : Z_0 x \geq EZ_T F, \quad \forall Z \in \mathcal{M}_0^T(G^*)\right\}.$$

From the last remark, we deduce that

$$\Gamma_F = G_0 + (0, 1).$$

In a same way, if we note

$$G_0^{*\perp} := \{x : \exists g \in G_0^* \text{ such that } x g = 0\},$$

we deduce that the set of all the fair prices associated to F is

$$I_F = G_0^{*\perp} + (0, 1).$$

From this, we deduce that $I_F \cap \Gamma_F = \Gamma_F^m$ as presented in the following graphic with $\lambda = 0.25$.

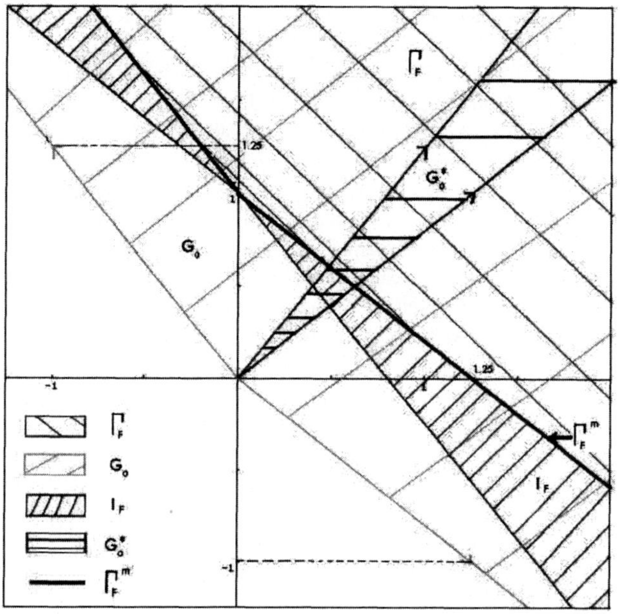

Note that we obtain an analogous result for a two-dimensional Y-model for which we suppose that $\mathcal{D} \neq \emptyset$, and for any $g \in G_0^*$, there exists a martingale $Z \in \mathcal{M}_0^T(G^*)$ such that $Z_0 = g$ and F is a contingent claim of the form $F = (a, b)$ where a, b are two constants. This is the case if the transaction costs coefficients verifies $\lambda_t^{i,j} \geq \lambda_0^{i,j} > 0$, $i, j = 1, 2$ and the price process is $(1, S)$ where $dS_t = \sigma S_t dW_t$, W being a standard brownian motion.

References

1. Bouchard, B. and Chassagneux, J. F., *Representation of continuous linear forms on the set of ladlag processes and the pricing of American claims under proportional costs*, (2008), Preprint.
2. Cherny, A., General arbitrage pricing model. I. Probability approach, *Séminaire de Probabilités XL, 415–445, Lecture Notes in Math.*, 1899, (2007), Springer, Berlin.
3. Cherny, A. General arbitrage pricing model. II. Transaction costs, *Séminaire de Probabilités XL, 447–461, Lecture Notes in Math.*, 1899, (2007), Springer, Berlin.
4. Guasoni, P., Rásony, M., and Schachermayer, W., The Fundamental Theorem of Asset Pricing for Continuous Processes under Small Transaction Costs, *Annals of Finance*, (2008).
5. Guasoni, P. and Rásony, M., *The Fundamental Theorem of Asset Pricing under Transaction Costs*, (2008), Working Paper.
6. Jouini, E. and Kallal, H., Martingales and Arbitrage in Securities Markets with Transaction Costs, *J. Econom. Theory*, **66**, (1995), 178–197.
7. Kabanov, Yu. and Safarian, M., *Markets with Transaction Costs. Mathematical Theory*, Monograph, To appear.

8. Kabanov, Yu., Hedging and Liquidation Under Transaction Costs in Currency Markets, *Finance and Stochastics*, **3**, (1999), 237–248.
9. Kabanov, Yu. and Stricker, Ch., The Harrison-Pliska Arbitrage Pricing Theorem under Transaction Costs, *J. Maths. Enonom*, **35**, (2002), 185–196.
10. Kabanov, Yu., Rásony, M., and Stricker, Ch., No Arbitrage Criteria for Financial Markets with Efficient Friction, *Finance and Stochastics*, **6**, (2002), 371–382.
11. Kabanov, Yu., Hedging Under Transaction Costs in Currency Markets: a Continuous-Time Model, *Finance and Stochastics*, **3**, (1999), 237–248.
12. Schachermayer, W., The Fundamental Theorem of Asset Pricing under Proportional Transaction Costs in Finite Discrete Time, *Math. Finance*, **14**, (2004), 19–48.
13. Denis, E., De Vallière, D., and Kabanov Yu., Hedging of American Options under Transaction Costs, *Finance and Stochastics*, **13**, (2008), 105–119.

Leland's Approximations for Concave Pay-off Functions

Emmanuel Denis

School of Management, Boston University
E-mail: emmanuel.denis6@wanadoo.fr, edenis@bu.edu

In 1985, Leland suggested an approach to pricing contingent claims under proportional transaction costs. Its main idea is to use the classical Black–Scholes formula with a suitably enlarged volatility for a periodically revised portfolio which terminal value approximates the pay-off $h(S_T)$. In subsequent studies, Lott (for $\alpha = 1/2$), Kabanov and Safarian proved that for the call-option, i.e. for $h(x) = (x-K)^+$, Leland's portfolios, indeed, approximate the pay-off if the transaction costs coefficients decreases as $n^{-\alpha}$ for $\alpha \in]0, 1/2]$ where n is the number of revisions. These results can be extended to the case of more general pay-off functions and non-uniform revision intervals [1]. Unfortunately, the terminal values of portfolios do not converge to the pay-off if h is not a convex function. In this paper, we show that we can slightly modify the Leland strategy such that the convergence holds for a large class of concave pay-off functions if $\alpha = 1/2$.
Key words: Black–Scholes formula, Transaction costs, Leland's strategy, Approximate hedging.

1. Introduction

In his famous paper [7] Leland suggested, in the framework of a two-asset model of financial market with proportional transaction costs, a modification of the Black–Scholes approach to pricing contingent claims. The idea is simple: one can use the Black–Scholes formula but not with a true volatility parameter σ but with an artificially enlarged one, $\widehat{\sigma}$. A theoretical justification of this approach is based on the replication principle: the terminal value of a "real-world" self-financing portfolio, revised at sufficiently large number n of dates t_k, should approximate the terminal pay-off. Leland gave an explicit formula for enlarged volatility $\widehat{\sigma}$ which may depend on n. His pricing methodology is of great practical importance, in particular, due to an easy implementation.

However, a mathematical validation of this "approximate replication principle" happened to be quite delicate. The first rigorous result was obtained by Lott [9] who shown that the convergence in probability, as was conjectured by Leland, holds when the transaction costs coefficient $k_n = k_0 n^{-\alpha}$ decreases to zero for $\alpha = 1/2$ (in this case, $\widehat{\sigma}$ does not depend on n). On the other hand, for the constant k_0 the replication principle fails to be true. This was observed by Kabanov and Safarian [5] who calculated the limiting approximation error. They also proved that the replication error tends to zero when $\alpha \in]0, 1/2[$. Interesting limit theorems for the case $\alpha = 0$ (i.e. constant k) were obtained by Granditz and Schachinger [3] and Pergamenshchikov [10]. Recent notes [12], [13] and [8] constitute a discussion on the magnitude of the replication error for practical values of prices and transaction costs coefficients. Results on the first-order asymptotics of the L^2-norm of the approximation error can be found in [2]. All mentioned papers deal with the call option, i.e. with the particular pay-off function $h(x) = (x - K)^+$. Even in this case the arguments need a lot of estimates. The explicit expressions given by the Black–Scholes formula simplify calculations which are quite involved.

In the paper [1], we establish convergence results for more general pay-off functions and non-uniform revision intervals following the methodology of [2]. In particular, we show, for the case $\alpha \in]0, 1/2]$, that the approximation error converges to zero for convex pay-off functions of "moderate" growth. For non-convex pay-off functions, a systematic error appears depending on the value of the stock price at maturity. We show in this paper that it's easy enough to get the convergence to the pay-off for concave functions in the case where the transaction costs decrease as $n^{-1/2}$, n being the number of revision dates tending to ∞.

2. Main Results

We consider the standard two-asset model with the time horizon $T = 1$ assuming that it is specified under the martingale measure, the non-risky asset is the *numéraire*, and the price of risky asset is given by the formula

$$S_t = S_0 \exp\left\{-\frac{1}{2}\int_0^t \sigma_s^2 ds + \int_0^t \sigma_s dW_s\right\}.$$

where W is a Wiener process. So, $dS_t = \sigma_t S_t dW_t$. We assume that $\sigma(t)$ is a strictly positive and continuous function on $[0, 1]$ verifying

$$|\sigma(t) - \sigma(u)| \leq \mu |t - u|$$

where $\mu > 0$ is a constant. In particular, we have $\sigma(t) \in [\sigma_1, \sigma_2]$ where $\sigma_1 > 0$. Note that

$$S_t \sim S_0 \exp\{\alpha_t G - \alpha_t^2/2\}$$

where $\alpha_t^2 = \int_0^t \sigma_s^2 ds$ and $G \sim \mathcal{N}(0, 1)$.
Recall that, according to Black and Scholes, the price of the contingent claim

$h(S_1)$ is the initial value of the replicating portfolio

$$V_t = V_0 + \int_0^t H_r dS_r = E(h(S_1)|\mathcal{F}_t) = C(t, S_t),$$

where

$$C(t,x) = E\, h(x \exp\{\overline{\rho}_t G - \overline{\rho}_t^2/2\}),$$
$$\overline{\rho}_t = \int_t^1 \sigma_s^2 ds$$

and the replication strategy is $H_r = C_x(r, S_r)$.

In the model with proportional transaction costs and a finite number of revisions the current value of the portfolio process at time t is described as

$$V_t^n = V_0^n + \int_0^t H_u^n dS_u - \sum_{t_i \le t} k_n S_{t_i} |H_{i+1}^n - H_i^n| \tag{1}$$

where H^n is a piecewise-constant process with $H^n = H_i^n$ on the interval $]t_{i-1}, t_i]$, $t_i = t_i^n$, $i \le n$, are the revision dates, and H_i^n are $\mathcal{F}_{t_{i-1}}$-measurable random variables. We assume that the transaction costs coefficient verifies

$$k = k_n = k_0 n^{-1/2}, \tag{2}$$

and the dates t_i are defined by a strictly increasing function $g \in C^1[0,1]$ with $g(0) = 0$, $g(1) = 1$, so that $t_i = g(\frac{i}{n})$. Let denote by f the inverse of g. The "enlarged volatility" is given by the formula

$$\widehat{\sigma}_t^2 = \sigma_t^2 - \sigma_t k_0 \sqrt{8/\pi}\sqrt{f'(t)} = \sigma_t^2 - \sigma_t \gamma(t) \tag{3}$$

where we assume that g is chosen such that there exists $\beta < 1/3$ verifying for n large enough

$$\frac{1}{k_0}\sqrt{\frac{\pi}{8}}\sigma_t - \sqrt{f'(t)} > \frac{1}{n^\beta}, \quad \forall t \in [0,1].$$

We call the Leland strategy the process H^n with

$$H_i^n = \widehat{C}_x(t_{i-1}, S_{t_i-1})$$

where the function $\widehat{C}(t,x)$ is the solution of the Cauchy problem on $[0,1]$:

$$\widehat{C}_t(t,x) + \frac{1}{2}\widehat{\sigma}_t^2 x^2 \widehat{C}_{xx}(t,x) = 0, \qquad \widehat{C}(1,x) = h(x). \tag{4}$$

Its solution can be written as

$$\widehat{C}(t,x) = \int_{-\infty}^{\infty} h(xe^{\rho y - \rho^2/2})\varphi(y)dy \tag{5}$$

where $\rho^2 = (\rho_t)^2 = \int_t^1 \widehat{\sigma}_s^2 ds$ and φ is the Gaussian density.

Note that $\widehat{\sigma}_s^2 \geq cn^{-\beta}$ for a constant $c > 0$ and n large enough and, therefore,

$$\rho^2 \geq cn^{-\beta}(1-t).$$

We shall use the following hypothesis on the "cadence" of revisions:

Assumption (G1): $g' > 0$ and $g \in C^2[0,1]$.

Assumption (G2): the function g is

$$g_\mu(t) = 1 - (1-t)^\mu,\ \mu \in]\beta/2, 1[.$$

It is easy to see that, in this two cases, the following properties hold:

Lemma 2.1. *Assume that (G1) or (G2) hold. Then, there exists some constants $d_1, d_2 > 0$ such that for $i = 0, .., n-1$ and n enough large:*

$$\Delta t_i = t_i - t_{i-1} \leq d_1 n^{-\mu},\ 1 - t_{n-1} \geq d_2 n^{-\mu} \tag{6}$$

$$\frac{1 - t_{i-1}}{1 - t_i} \leq d_1, \tag{7}$$

$$u \in [t_{i-1}, t_i[\Rightarrow f'(u)(t_i - u)n \leq d_1. \tag{8}$$

We use the abbreviations $\widehat{H}_t = \widehat{C}_x(t, S_t)$ and $\widehat{h}_t = \widehat{C}_{xx}(t, S_t)$.

Our hypothesis on the pay-off function is as follows:

Assumption (H): h is a continuous function on $[0, \infty[$ which is two-times differentiable except the points $K_1 < \cdots < K_p$ where h' and h'' admit right and left limits; $|h''(x)| \leq Mx^{-\beta}$ for $x \geq K_p$ where $\beta \geq 3/2$.

Let $K_0 = 0$ and $K_{p+1} = \infty$. Then h', h'' are bounded while h verifies the inequality $|h(x)| \leq M(1+x)$ for a constant M. The function $\widehat{C}(t,x)$ is continuous on $[0,1] \times \mathbf{R}$.

Theorem 2.1. *Let $k_n = k_0 n^{-1/2}$ where $k_0 > 0$. Suppose that Assumptions (G1) or (G2) and (H) hold. Then*

$$P\text{-}\lim_n V_1^n = h(S_1). \tag{9}$$

3. Estimates

In the following subsections we recall from [1] some properties of the solution of the Cauchy problem (4) needed for the proof of Theorem 2.1.

3.1 Explicit Formulae

Recall the following formulae for the derivatives we can find in [1].

Lemma 3.1. *Let $\widehat{C}(t,x)$ is given by (5). Then*

$$\frac{\partial^{k+1}\widehat{C}(t,x)}{\partial x^{k+1}} = \frac{1}{\rho^k x^k}\int_{-\infty}^{\infty} h'(xe^{\rho y+\rho^2/2})P_k(y)\varphi(y)dy, \qquad k \geq 0,$$

where $P_k(y) = y^k + a_{k-1}(\rho)y^{k-1} + \cdots + a_0(\rho)$ is a polynomial of degree k whose coefficients $a_i(\rho)$ are polynomials in ρ of degree $k-1$.

In particular, $|\widehat{C}_x(t,x)| \leq \|h'\|_\infty$ and we have the following expressions:

(1) $$\widehat{C}_{xx}(t,x) = \frac{1}{\rho x}\int_{-\infty}^{\infty} h'(\tilde{z}(y,x,\rho))y\varphi(y)dy,$$

(2) $$\widehat{C}_{xxx}(t,x) = \frac{1}{\rho^2 x^2}\int_{-\infty}^{\infty} h'(\tilde{z}(y,x,\rho))(y^2-\rho y-1)\varphi(y)dy,$$

(3) $$\widehat{C}_{xxxx}(t,x) = \frac{1}{\rho^3 x^3}\int_{-\infty}^{\infty} h'(\tilde{z}(y,x,\rho))P_3(y)\varphi(y)dy$$

with $P_3(y) = y^3 - 3\rho y^2 + (2\rho^2-3)y + 3\rho$.

By similar reasoning we obtain:

Lemma 3.2. *Let $\widehat{C}(t,x)$ is given by (5). Then*

(4) $$\widehat{C}_t(t,x) = \frac{-\widehat{\sigma}_t^2 x}{2\rho}\int_{-\infty}^{\infty} h'(xe^{\rho y+\rho^2/2})y\varphi(y)dy,$$

(5) $$\widehat{C}_{tx}(t,x) = \frac{\widehat{\sigma}_t^2}{2\rho^2}\int_{-\infty}^{\infty} h'(xe^{\rho y+\rho^2/2})(-y^2-\rho y+1)\varphi(y)dy,$$

(6) $$\widehat{C}_{xxt}(t,x) = \frac{\widehat{\sigma}_t^2}{2\rho^3 x}\int_{-\infty}^{\infty} h'(xe^{\rho y+\rho^2/2})Q_3(y)\varphi(y)dy,$$

with $Q_3(y) = -y^3 - \rho y^2 + 3y + \rho$.

3.2 Inequalities

From [1], we have the following estimates:

Lemma 3.3. *There exists a constant c such that*

$$|\widehat{C}_{xxx}(t,x)| \leq \frac{ce^{-\rho^2/8}}{\rho^2 x^{5/2}}\left(L(x,\rho)+\rho\right),$$

$$|\widehat{C}_{xxxx}(t,x)| \leq ce^{-\rho^2/8}x^{-7/2}P_3(\rho^{-1}),$$

$$|\widehat{C}_{tx}(t,x)| \leq \frac{c\widehat{\sigma}^2 e^{-\frac{\rho^2}{8}}}{x^{1/2}\rho^2}\left(L(x,\rho)+\rho+\rho^2\right),$$

$$|\widehat{C}_{xxt}(t,x)| \leq c\widehat{\sigma}^2 e^{-\rho^2/8}x^{-3/2}(\rho^{-1}+\rho^{-3}),$$

where P_3 is a polynomial of the third order and

$$L(x,\rho) = \sum_{j=1}^{p} \frac{|\ln(x/K_j)|}{\rho} \exp\left\{-\frac{\ln^2(x/K_j)}{2\rho^2}\right\}.$$

Corollary 3.1. *There exists a constant c such that for $t \in [0,1[$*

$$ES_t^4 \widehat{C}_{xx}^2(t,S_t) \leq c\frac{1}{\rho}e^{-\rho^2/4}.$$

4. Proof of Theorem 2.1

By the Ito formula we get that

$$\widehat{C}_x(t,S_t) = \widehat{C}_x(0,S_0) + M_t^n + A_t^n \tag{1}$$

where

$$M_t^n := \int_0^t \sigma_u S_u \widehat{C}_{xx}(u,S_u)dW_u,$$

$$A_t^n := \int_0^t \left[\widehat{C}_{xt}(u,S_u) + \frac{1}{2}\sigma_u^2 S_u^2 \widehat{C}_{xxx}(u,S_u)\right]du.$$

The process M^n is a square integrable martingale on $[0,1]$ in virtue of [1].

Following [6] we represent the difference $V_1^n - h(S_1)$ in a convenient form.

Lemma 4.1. *We have $V_1^n - h(S_1) = F_1^n + F_2^n$ where*

$$F_1^n := \int_0^1 (H_t^n - \widehat{H}_t)dS_t - k_n|H_{t_n}^n - H_{t_{n-1}}^n|S_{t_n}, \tag{2}$$

$$F_2^n := -\frac{1}{2}\int_0^1 \sigma_t \gamma_n(t) S_t^2 \widehat{C}_{xx}(t,S_t)dt - k_n \sum_{i=1}^{n-1} |H_{t_i}^n - H_{t_{i-1}}^n|S_{t_i}. \tag{3}$$

Put $D_i^a := \{(x,y) : x \in [1/a,a],\ xe^y \in]K_i,K_{i+1}[\}$, $a>0$, $i=0,\dots,p$. Recall the following lemma [1]:

Lemma 4.2. *The mapping $(x,y) \mapsto h'(xe^y)$ is a Lipschitz function on each set D_i^a, i.e. there exists a constant L_a such that*

$$|h'(xe^y) - h'(ze^u)| \leq L_a(|x-z| + |y-u|)$$

for all $x,z \in [1/a,a]$ and y, u such that xe^y, $ze^u \in]K_i,K_{i+1}[$, $i=0,\dots,p$.

Then, we deduce the following:

Lemma 4.3.

$$P\text{-}\lim_n F_1^n = 0. \tag{4}$$

Proof.

The proof is clearly the same as in [1] because we have the following inequality $|\rho_u^n - \rho_{t_{i-1}}^n| \le c_u n^{\beta/2-\mu}$ for some constant c_u and $u \in [t_{i-1}, t_i]$. Then, it suffices to use the previous lemma. □

We write $F_2^n = \sum_{i=1}^5 L_i^n$ where

$$L_1^n := -\frac{1}{2}\int_0^1 \sigma_t \gamma(t) S_t^2 \widehat{h}_t dt + \frac{1}{2}\int_0^1 \sum_{i=1}^{n-1} \sigma_{t_{i-1}} \gamma(t_{i-1}) S_{t_{i-1}}^2 \widehat{h}_{t_{i-1}} I_{]t_{i-1},t_i]}(t) dt$$

$$L_2^n := -\sum_{i=1}^{n-1} \widehat{h}_{t_{i-1}} S_{t_{i-1}}^2 \left(\frac{1}{2}\sigma_{t_{i-1}} \gamma(t_{i-1}) \Delta t_i - k_n \sigma_{t_{i-1}} n^{1/2} \sqrt{\Delta t_i f'(t_{i-1})} |\Delta W_{t_i}| \right),$$

$$L_3^n := -k_n \sum_{i=1}^{n-1} \sigma_{t_{i-1}} S_{t_{i-1}}^2 \widehat{h}_{t_{i-1}} n^{1/2} \sqrt{\Delta t_i f'(t_{i-1}} |\Delta W_{t_i}| - k_n \sum_{i=1}^{n-1} S_{t_{i-1}} |\Delta M_{t_i}|,$$

$$L_4^n := k_n \sum_{i=1}^{n-1} S_{t_{i-1}} |\Delta M_{t_i}| - k_n \sum_{i=1}^{n-1} S_{t_{i-1}} |\Delta \widehat{H}_{t_i}|,$$

$$L_5^n := -k_n \sum_{i=1}^{n-1} \Delta S_{t_i} |\Delta \widehat{H}_{t_i}|$$

where we use the abbreviations $\Delta W_{t_i} = W_{t_i} - W_{t_{i-1}}$ etc.

Lemma 4.4. *Both terms whose difference defines L_1^n converge almost surely, as $n \to \infty$, to the random variable J given by the formulae*

$$J = -\frac{1}{2}\int_0^1 \sigma_t \gamma(t) S_t^2 \widehat{h}_t dt. \tag{5}$$

Therefore, $L_1^n \to 0$ a.s.

Proof. The first term is a constant since the function $\gamma(t)$ and the solution of the Cauchy problem don't depend on n and $L_1^n \to 0$ because of convergence of the Riemann sums to the integral. □

Lemma 4.5. *We have P-$\lim_n L_2^n = 0$.*

Proof. Taking into account the independence of increments of the Wiener process and the equalities

$$E|\Delta W_{t_i}| = \sqrt{2/\pi}\sqrt{\Delta t_i},$$

$$E\left(\frac{1}{2}\gamma(t_{i-1})\Delta t_i - k_n \sqrt{n \Delta t_i f'(t_{i-1})} |\Delta W_{t_i}| \right)^2 = (1 - 2/\pi) k_n^2 n f'(t_{i-1}) (\Delta t_i)^2,$$

we obtain that

$$E(L_2^n)^2 = (1-2/\pi)k_n^2 \sum_{i=1}^{n-1} \sigma_{t_{i-1}}^2 ES_{t_{i-1}}^4 \widehat{C}_{xx}^2(t_{i-1}, S_{t_{i-1}}) f'(t_{i-1})\, n(\Delta t_i)^2$$

where $f'(t_{i-1})\Delta t_i\, n$ is bounded. In virtue of Lemma 3.1

$$E(L_2^n)^2 \le c\, k_0^2\, n^{\beta/2-1} \sum_{i=1}^{n-1} (1-t_{i-1})^{-1/2} \Delta t_i$$

and so $E(L_2^n)^2 \to 0$ as $n \to \infty$. □

Lemma 4.6. *For any $\alpha \in [0, 1/2]$, we have P-$\lim_n L_3^n = 0$.*

Proof. In the case of the assumption $(G1)$, we use the approximation

$$n^{1/2} \sqrt{f'(t_{i-1})\Delta t_i} = 1 + \varepsilon_n$$

where $\varepsilon_n = O_n(n^{-1})$. We deduce that $L_3^n = C^n + D^n$ with

$$C^n := k_n \varepsilon_n \sum_{i=1}^{n-1} \sigma_{t_{i-1}} S_{t_{i-1}}^2 |\widehat{h}_{t_{i-1}}| \, |\Delta W_{t_i}|$$

and

$$D^n := k_n \sum_{i=1}^{n-1} \left[\sigma_{t_{i-1}} S_{t_{i-1}}^2 |\widehat{h}_{t_{i-1}}| \, |\Delta W_{t_i}| - S_{t_{i-1}} |\Delta M_{t_i}| \right].$$

Using Lemma 3.1 we obtain:

$$\|C^n\|_2 \le k_n \varepsilon_n \sum_{i=1}^{n-1} \frac{n^{\beta/4}}{(1-t_{i-1})^{1/4}} \sqrt{\Delta t_i},$$

$$\|C^n\|_2 \le const\, \varepsilon_n\, n^{\beta/4} \int_0^1 \frac{dt}{(1-t)^{1/4}}$$

proving that $\|C^n\|_2 \to 0$ as $n \to \infty$.

For Assumption $(G2)$, we write

$$n^{1/2} \sqrt{f'(t_{i-1})\Delta t_i} = 1 + \varepsilon_i$$

where

$$|\varepsilon_i| \le c \frac{\Delta t_i}{1 - t_i}$$

for a constant c because of the assumption. Thus, we have

$$\|C^n\|_2 \le c\,n^{-1/2}\sum_{i=1}^{n-1}\frac{n^{\beta/4}(\Delta t_i)^{3/2}}{(1-t_i)^{5/4}} \le c\,n^{\beta/4-1/2}\ln n.$$

The remaining part is similar to the proof in [1]: $|D^n| \le D_1^n + D_2^n$ where

$$D_1^n := k_n\sum_{i=1}^{n-1} S_{t_{i-1}}\left|\int_{t_{i-1}}^{t_i}(S_{t_{i-1}}\widehat{h}_{t_{i-1}} - S_u\widehat{h}_u)\sigma_u dW_u\right|,$$

$$D_2^n := k_n\sum_{i=1}^{n-1}\left|\int_{t_{i-1}}^{t_i}(\sigma_{t_{i-1}} - \sigma_u)S_{t_{i-1}}^2\widehat{h}_{t_{i-1}}dW_u\right|.$$

We have obviously $\|D_2^n\|_2 \to 0$ whereas

$$E|D_1^n| \le ck_n\sum_{i=1}^{n-1}\left(\int_{t_{i-1}}^{t_i}E(S_{t_{i-1}}\widehat{h}_{t_{i-1}} - S_u\widehat{h}_u)^2du\right)^{1/2}.$$

By the Ito formula

$$d[S_t\widehat{h}_t] = d[S_t\widehat{C}_{xx}(t,S_t)] = f_tdW_t + g_tdt$$

where

$$f_t := \sigma_tS_t\widehat{C}_{xx}(t,S_t) + \sigma_tS_t^2\widehat{C}_{xxx}(t,S_t) = \frac{\sigma_t}{\rho^2}\int_{-\infty}^{\infty}h'(\widetilde{z})(y^2-1)\varphi(y)dy,$$

$$g_t := S_t\widehat{C}_{xxt}(t,S_t) + \frac{1}{2}\sigma_t^2S_t^3\widehat{C}_{xxxx}(t,S_t) + \sigma_t^2S_t^2\widehat{C}_{xxx}(t,S_t).$$

Thus,

$$E(S_{t_{i-1}}\widehat{h}_{t_{i-1}} - S_t\widehat{h}_t)^2 \le 2\int_{t_{i-1}}^{t_i}Ef_u^2du + 2\Delta t_i\int_{t_{i-1}}^{t_i}Eg_u^2du.$$

It follows from Lemma 3.3 that

$$\begin{aligned}E|D_1^n| &\le c\,n^{\beta-1/2}\sum_{i=1}^{n-1}\frac{\Delta t_i}{1-t_i} + c\,n^{3\beta/2-1/2}\sum_{i=1}^{n-1}\frac{(\Delta t_i)^{3/2}}{(1-t_i)^{3/2}}\\ &\le c\,n^{\beta-1/2}\ln n + c\,n^{3\beta/2-1/2}\ln n.\end{aligned}$$

Thus, $E|D_1^n|$ converge to zero. □

Lemma 4.7. *We have* $P\text{-}\lim_n L_4^n = 0$.

Proof. Using again the inequality $||a_1| - |a_2|| \le |a_1 - a_2|$ we get that

$$|L_n^4| \le k_n \sum_{i=1}^{n-1} S_{t_{i-1}} |\Delta A_{t_i}|$$
$$\le c(\omega) k_n \int_0^{t_{n-1}} |\widehat{C}_{xt}(u, S_u)| du + c(\omega) k_n \int_0^{t_{n-1}} \sigma_u^2 S_u^2 |\widehat{C}_{xxx}(u, S_u)| du.$$

Estimating the first integral using Lemmas 3.3 and 2.1 we obtain that

$$k_n \int_0^{t_{n-1}} |\widehat{C}_{xt}(u, S_u)| du \le c\, n^{\beta - 1/2} \ln n \tag{6}$$

which implies the convergence of the latter to 0.

The reasoning is similar to analyze the second term. □

Lemma 4.8. *We have* $P\text{-}\lim_n L_5^n = 0$.

Proof. Since $\max_i |\Delta S_{t_i}| \to 0$ as $n \to \infty$, it suffices to verify that the sequence $k_n \sum_{i=1}^n |\Delta \widehat{H}_{t_i}|$ is bounded in probability. But this follows from the preceding lemmas. □

Inspecting the formulations of above lemmas, we observe that all terms $L_j^n \to 0$ in probability and, hence, Theorem 2.1 is proven.

References

1. Emmanuel, D., Approximate Hedging of Contingent Claims under Transaction Costs for General Pay-off, submitted to *Applied Mathematical Finance.*
2. Gamys M. and Kabanov Yu. M., Mean square error for the Leland–Lott hedging strategy, submitted to *Finance and Stochastics.*
3. Granditz P. and Schachinger W., Leland's approach to option pricing: The evolution of discontinuity, *Mathematical Finance*, **11**, (2001), 3, 347–355.
4. Grannan E. R. and Swindle G. H., Minimizing transaction costs of hedging strategies. *Mathematical Finance*, **6**, (1996), 4, 341–364.
5. Kabanov Yu. M. and Safarian M. On Leland's strategy of option pricing with transaction costs. *Finance and Stochastics*, **1**, 3, 1997, 239–250.
6. Kabanov Yu. and Safarian M. Markets with Transaction Costs. Mathematical Theory. Monograph. To appear.
7. Leland H. Option pricing and replication with transactions costs, *Journal of Finance*, **XL** (1985), 5, 1283–1301.
8. Leland H. Comments on "Hedging errors with Leland's option model in the presence of transactions costs". *Finance Research Letters*, **4** (2007), 3, 200–202.
9. Lott K. Ein Verfahren zur Replikation von Optionen unter Transaktionkosten in stetiger Zeit, Dissertation. Universität der Bundeswehr München. Institut für Mathematik und Datenverarbeitung, 1993.
10. Pergamenshchikov S. Limit theorem for Leland's strategy. *Annals of Applied Probability*, **13** (2003), 1099–1118.

11. Sekine J. and Yano J. Hedging errors of Leland's strategies with time-inhomogeneous rebalancing. Preprint.
12. Zhao Y. and Ziemba W. T. Hedging errors with Leland's option model in the presence of transaction costs. *Finance Research Letters*, **4** (2007), 1, 49–58.
13. Zhao Y. and Ziemba W. T. Comments on and corrigendum to "Hedging errors with Leland's option model in the presence of transaction costs". *Finance Research Letters*, **4** (2007), 3, 196–199.

Option Pricing Based on Geometric Stable Processes and Minimal Entropy Martingale Measures

Yoshio Miyahara* and Naruhiko Moriwaki†

* Graduate School of Economics, Nagoya City University
E-mail: y-miya@econ.nagoya-cu.ac.jp
† MTEC (Mitsubishi UFJ Trust Investment Technology Institute Co., Ltd.)

The objective of this paper is to introduce the [GSP & MEMM] (geometric stable processes and minimal entropy martingale measure) pricing model and show its advantages for pricing options with its fat tailed property. As the result of empirical analysis the good fitness of the [GSP & MEMM] pricing model is shown to the currency options.
Key words: incomplete market, option pricing, Lévy process, stable process, fat tailed property, minimal entropy martingale measure, calibration

1. Introduction

It is well-known that the distribution of the log returns from the market data of the asset prices have the asymmetry and the fat tail property. And the importance of the stable process was discussed in Fama [8]. The option pricing models based on the stable process have been studied by many researchers (Edelmann [6], Rachev and Mittnik [21], Carr and Wu [3], etc.). But in their models the martingale measures were not clear.

The minimal entropy martingale measure (MEMM=CMM) was introduced in Miyahara [15] and the fundamental properties of MEMM were investigated in it. Frittelli [9] showed that the MEMM is related to the exponential utility function. The MEMM for the geometric Lévy process (GLP) has been studied by Fujiwara and Miyahara [10], Esche and Schweizer [7], Hubalek-Sgarra [11], etc.

Combining the GLP and the MEMM, we obtain the [GLP & MEMM] pricing models (Miyahara [17]), and it is known that this model have many good properties as an option pricing model for the incomplete market (Miyahara [18]).

We put our focus mainly on the geometric stable processes (GSP), and we construct the [GSP & MEMM] (geometric stable process and minimal entropy martingale measure) model. We remark here that the MEMM is the only-one reasonable martingale measure the existence of which is confirmed for the GSP.

In section 2 we give the definition of the [GLP & MEMM] models and especially the [GSP & MEMM] model. In section 3 we explain the reproducibility of the volatility smile and smirk properties of these models. In section 4 we explain the fitness of the [GSP & MEMM] pricing models to the currency options. As the results of above studies, we could have proved the efficiency of the [GSP & MEMM] model.

2. [GLP & MEMM] and [GSP & MEMM] Models

The [GLP & MEMM] pricing model is the following model:

(A) The price process S_t of the underlying asset is a geometric Lévy process (GLP).
(B) The price of an option X is defined by $e^{-rT}E_{P^*}[X]$, where P^* is the MEMM for S_t. Where r is the interest rate of the risk free asset, and T is the maturity.

This model was first introduced in Miyahara [17], and the properties of this model are summarized in Miyahara [18], etc. Adopting the stable process for the Lévy process in (A), we obtain the [GSP & MEMM] Model.

2.1 Geometric Lévy Process (GLP)

The price process S_t of a stock is assumed to be defined as what follows. We suppose that a probability space $(\Omega, \mathcal{F}, P)$ and a filtration $\{\mathcal{F}_t, 0 \leq t \leq T\}$ are given, and that the price process S_t of a stock is defined on this probability space and given in the form

$$S_t = S_0 e^{Z_t}, \quad 0 \leq t \leq T, \tag{2.1}$$

where Z_t is a Lévy process. We call such a process S_t the geometric Lévy precess (GLP), and we denote the generating triplet of Z_t by $(\sigma^2, \nu(dx), b)$.

We see examples of Lévy processes.

2.1.1 Variance Gamma Process

The Lévy measure $\nu(dx)$ of a variance Gamma process is

$$\nu(dx) = C\frac{e^{-c_1|x|}I_{\{x<0\}} + e^{-c_2|x|}I_{\{x>0\}}}{|x|}dx, \ C, c_1, c_2 > 0,$$

and $\sigma^2 = 0$: The characteristic function is

$$\phi(u) = e^{ib_0 u}\left(\frac{1}{(1 + iu/c_1)(1 - iu/c_2)}\right)^C.$$

(Madan and Seneta [13])

2.1.2 CGMY Process

The Lévy measure of a CGMY process is

$$\nu(dx) = C\,\frac{I_{\{x<0\}}e^{-G|x|} + I_{\{x>0\}}e^{-M|x|}}{|x|^{1+Y}}\,dx,\ \ C, G, M > 0,\ Y < 2$$

and the characteristic function is

$$\begin{aligned}\phi(u) = \exp\Big\{&i\left(b_1 + C\Gamma(-Y)Y\left(M^{Y-1} - G^{Y-1}\right)\right)u \\ &+C\Gamma(-Y)\left((M - iu)^Y - M^Y + (G + iu)^Y - G^Y\right)\Big\},\end{aligned} \tag{2.2}$$

where $\Gamma(Y)$ is the Gamma function. (Carr, Geman, Madan and Yor [2])

2.1.3 Stable Model

The Lévy measure of a stable process is

$$\nu(dx) = \frac{c_1 I_{\{x<0\}} + c_2 I_{\{x>0\}}}{|x|^{1+\alpha}}dx,\ \ 0 < \alpha < 2,\ c_1, c_2 \geq 0,\ c_1 + c_2 > 0.$$

and the characteristic function is of the following form

$$\phi(u) = \exp(\psi(u))$$

$$\psi(u) = \begin{cases} \Gamma(-\alpha)\left(\cos\dfrac{\pi\alpha}{2}\right)c|u|^{\alpha}\left(1 - i\beta\tan\dfrac{\pi\alpha}{2}\mathrm{sgn}(u)\right) + i\tau u, & \text{for } \alpha \neq 1 \\ -\dfrac{\pi}{2}c|u|\left(1 + i\beta\dfrac{2}{\pi}\mathrm{sgn}(u)\log|u|\right) + i\tau u, & \text{for } \alpha = 1 \end{cases}$$

where

$$\begin{aligned} c &= c_1 + c_2, \\ \beta &= \frac{c_2 - c_1}{c_1 + c_2}, \\ \tau &= \begin{cases} \dfrac{c_2 - c_1}{1 - \alpha} - b, & \text{if } 0 < \alpha < 1 \\ b + c_0 c, & \text{if } \alpha = 1 \\ \dfrac{c_1 - c_2}{1 - \alpha} + b, & \text{if } 1 < \alpha < 2 \end{cases} \end{aligned}$$

$$c_0 = \int_0^1 r^{-2}(\sin r - r)\,dr + \int_1^{\infty} r^{-2}\sin r\,dr$$

2.1.4 Merton Jump-Diffusion Model

The process is

$$Z_t = b_0 t + \sigma W_t + J_t, \tag{2.3}$$

where J_t is the jump part and its Lévy measure is

$$\nu(dx) = \lambda \frac{1}{\sqrt{2\pi v}} \exp\left\{-\frac{(x-m)^2}{2v}\right\} dx, \quad \lambda > 0,\ v > 0. \tag{2.4}$$

The characteristic function is

$$\phi(u) = e^{\psi(u)}, \quad \psi(u) = iub_0 - \sigma^2 u^2/2 + \lambda\left\{\exp(ium - vu^2/2) - 1\right\} \tag{2.5}$$

(Merton [14])

2.1.5 NIG Process

The Lévy measure of a NIG (normal inverse Gaussian) process is

$$\nu(dx) = \frac{\delta\alpha}{\pi} \frac{\exp(\beta x) K_1(\alpha|x|)}{|x|} dx, \tag{2.6}$$

where $K_1(x)$ is the modified Bessel function of the third kind with index 1, and the characteristic function is

$$\phi(u) = \exp\left[ib_0 u + \delta\left(\sqrt{\alpha^2 - \beta^2} - \sqrt{\alpha^2 - (\beta + iu)^2}\right)\right]. \tag{2.7}$$

(Barndorff and Nielsen [1])

2.2 Minimal Entropy Martingale Measure (MEMM)

Next we will give the definition of the MEMM.

Definition 2.1. [minimal entropy martingale measure (MEMM)] If an equivalent martingale measure P^* satisfies

$$H(P^*|P) \leq H(Q|P) \qquad \forall Q: \ \textit{equivalent martingale measure}, \tag{2.8}$$

then P^* is called the minimal entropy martingale measure (MEMM) of S_t. Where $H(Q|P)$ is the relative entropy of Q with respect to P

$$(2.9) \qquad H(Q|P) = \begin{Bmatrix} \int_\Omega \log[\frac{dQ}{dP}]dQ, & if \quad Q \ll P, \\ \infty, & otherwise, \end{Bmatrix}.$$

For the existence of the MEMM, the following result is essential.

Theorem 2.1. ([10] Theorem 3.1) *Suppose that the following condition* **(C)** *holds*
Condition (C): *There exists $\theta^* \in R$ which satisfies the following conditions:*

$$(2.10) \qquad \mathbf{(C)_1} \quad \int_{\{x>1\}} e^x e^{\theta^*(e^x-1)} \nu(dx) < \infty,$$

$$\mathbf{(C)_2} \quad b + (\frac{1}{2} + \theta^*)\sigma^2 + \int_{\{|x|>1\}} (e^x - 1)e^{\theta^*(e^x-1)}\, \nu(dx)$$

$$(2.11) \qquad + \int_{\{|x|\le 1\}} \left((e^x - 1)e^{\theta^*(e^x-1)} - x\right) \nu(dx) = r_d - r_f.$$

Then the MEMM P^* exists, and (Z_t, P^*) is a Lévy process whose generating triplet (A^*, ν^*, b^*) is

$$A^* = \sigma^2,$$
$$\nu^*(dx) = e^{\theta^*(e^x-1)}\nu(dx),$$
$$b^* = b + \theta^*\sigma^2 + \int_{R\setminus\{0\}} xI_{\{|x|\le 1\}} d(\nu^* - \nu).$$

2.3 [GSP & MEMM] Model

When a class of Lévy process is selected, if the existence of the MEMM then a [GLP & MEMM] model is constructed. The properties of the [GLP & MEMM] models are described in Miyahara [17], [18]), etc. In the followings we restrict our focus on the [GSP & MEMM] model.

Applying the Theorem 1 to the geometric stable processes, we obtain the following the existence theorem for the MEMM.

Proposition 2.1. *For the geometric stable process, if $c_1, c_2 \ge 0$ and $c_1 + c_2 > 0$, then the equation* (2.11) *has a negative solution ($\theta^* < 0$) and the MEMM exists.*

So, the [GSP & MEMM] model could be constructed in general.

In the similar way, we obtain several kinds of [GLP & MEMM] Models.

3. Reproducibility of the Volatility Smile/Smirk Properties [1]

In this section, by the use of the simulation method, we see that [GLP & MEMM] models have the reproducibility of the volatility smile/smirk properties.

[1] The results of this section are based on the results of Miyahara-Moriwaki [19]. For the numerical caluculation we use the FFT(first Fourier transformation)-method and MATLAB-software.

3.1 Volatility Smile/Smirk Properties

Under the setting of the B-S (Black-Scholes) model, the historical volatility of the process is defined as the estimated value of the volatility σ based on the sequential data of the price process S_t. We denote it by $\widehat{\sigma}$. On the other hand the implied volatility is defined as what follows. Suppose that the market price of the European call option with the strike K, say $C_K^{(m)}$, were given. Then the value of σ, under which the theoretical B-S price of calloption is equal to $C_K^{(m)}$s, is the implied volatility, and this value is denoted by $\sigma_K^{(im)}$. We remark here that the implied volatility $\sigma_K^{(im)}$ depends on the strike value K.

If the market obeys to the B-S model, then the implied volatility $\sigma_K^{(im)}$ should be equal to the historical volatility $\widehat{\sigma}$. But in the real world this is not true. It is well-known that the implied volatility is not equal to the historical volatility, and the implied volatility $\sigma_K^{(im)}$ is sometimes a convex function of K, and sometimes the combination of convex part and concave part. These properties are so-called volatility smile or smirk properties. (See Cont and Fonseca [5] and Carr and Wu [4], etc.)

3.2 [Geometric Variance Gamma Process & MEMM] Model

The first example is the [Geometric Variance Gamma Process & MEMM] model. In the Figure 1, K is the strike price and S_0 is the prise of the underlying asset at present.

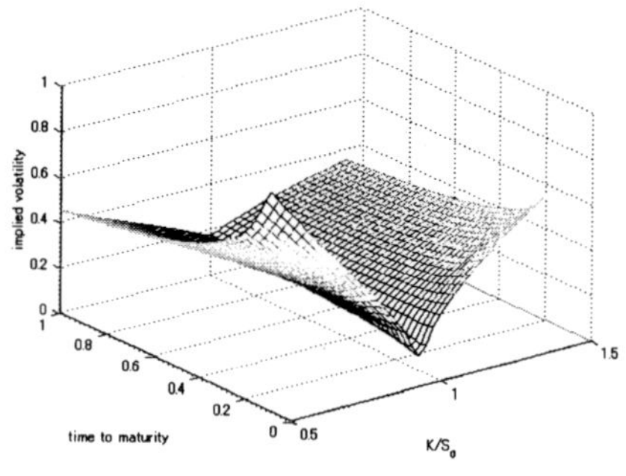

Figure 1. Implied volatility surface of the [Geometric VG Process & MEMM] Model

From this figure we can see that the [Geometric VG Process & MEMM] Model could reproduce the volatility smile property.

3.3 [Geometric CGMY Process & MEMM] Model

The next example is the [Geometric CGMY Process & MEMM] model.

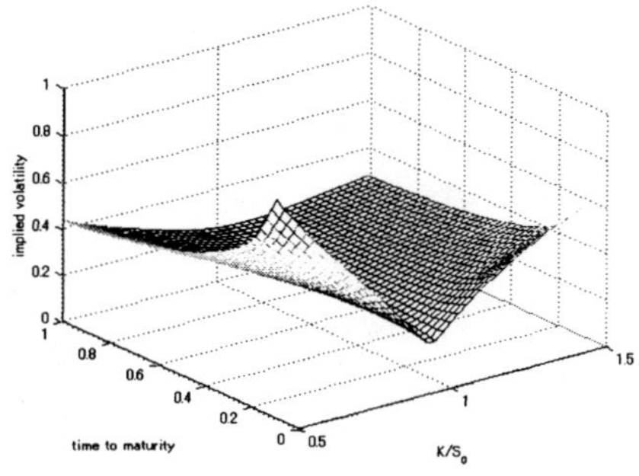

(a) volatility smile

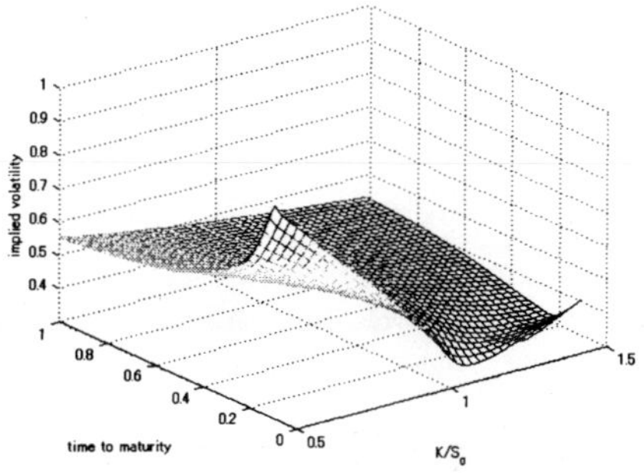

(b) volatility smirk

Figure 2. Implied volatility surface of [Geometric CGMY Process & MEMM] Model

From these figures we can see that the [Geometric CGMY Process & MEMM] Model could reproduce the volatility smile and volatility smirk properties.

3.4 [Geometric Stable Process & MEMM] Model

The next example is the [Geometric Stable Process & MEMM] model. In the figures, moneyness is defined as $moneyness = K/S_0$.

The above figures show us that the [Geometric Stable Process & MEMM] Model could reproduce the volatility smile and volatility smirk properties.

Moreover, changing parameters of this model, we can see that the [Geometric Stable Process & MEMM] Model could reproduce many kinds of volatility smile

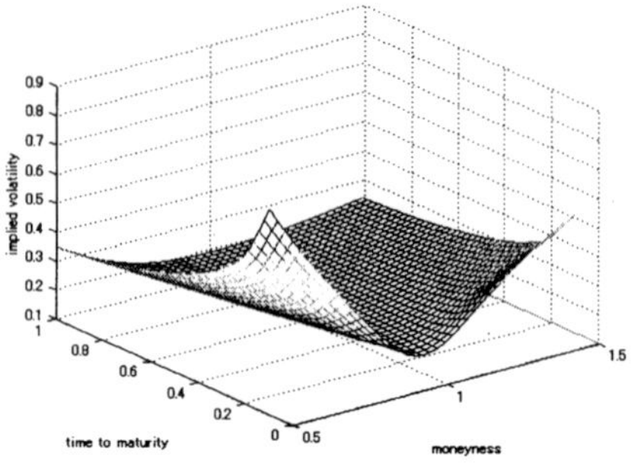

(a) volatility smile

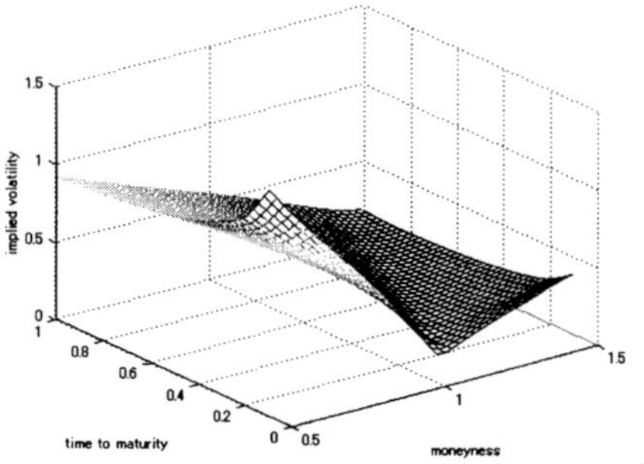

(b) volatility smirk

Figure 3. Implied volatility surface of [Geometric Stable Process & MEMM] Model

and volatility smirk surfaces. (See Miyahara and Moriwaki [19] for the details.)

And summing up the results of Miyahara and Moriwaki [19], we can obtain the following figures.

The above figures show us that the [Geometric Stable Process & MEMM] Model has the reproducibility of many types of volatility smile and smirk properties. We remark here that the strong smirk appears when the strong fat tail exists on right side.

Remark 3.1. Relating to the stable process, the FMLS(finite moment log stable) model is introduced in Carr and Wu [3]. This model is the special case of the GSP model, namely such restricted cases that $c_2 = 0$ in the GSP model. Introducing the MEMM, we can construct a little more extended models than FMLS models.

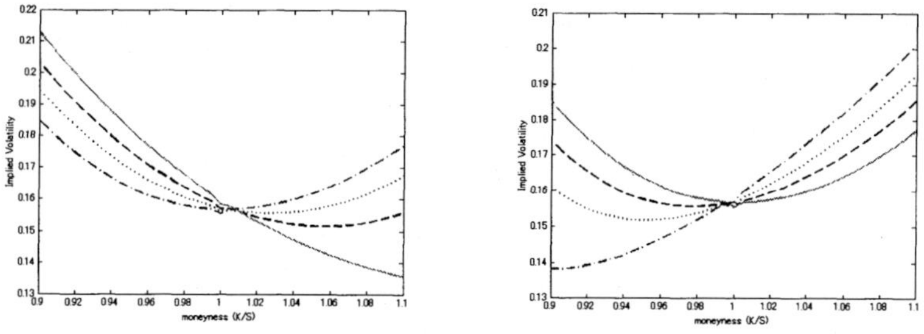

Figure 4. Volatility smile/smirk properties of the [G-Stable & MEMM] model. $b = 0$, $\alpha = 1.7$ and $r_d = r_f = 0, S = 1, T = 0.25$. Left Figure : dash-dot line ($c_1 = 0.005, c_2 = 0.005$), dotted line ($c_1 = 0.0065, c_2 = 0.0035$), dashed line ($c_1 = 0.008, c_2 = 0.002$), solid line ($c_1 = 0.01, c_2 = 1.0e - 11$). Right Figure : dash-dot line ($c_1 = 1.0e - 11, c_2 = 0.01$), dotted line ($c_1 = 0.002, c_2 = 0.008$), dashed line ($c_1 = 0.0035, c_2 = 0.0065$), solid line ($c_1 = 0.005, c_2 = 0.005$).

4. Application to Dollar-Yen Currency Options [2]

It is well-known that an exchange rate process has a strong fat tails and volatility smirk property. In fact we have obtained the following figure for Dollar-Yen currency options.

The results of the previous sections suggest us that the [GSP & MEMM] model could be adjusted to the currency option pricing.

4.1 Pricing Error for Option Price

Sometimes we estimate the option pricing models from the market data of option prices. This is the calibration.

Suppose that we have selected one type of the [GLP & MEMM] Model and that the parameters of the process of the underlying asset (or index) have been estimated. We denote the theoretical price of European call options based on the estimated process by $\widehat{C^*}(K, T)$, where K is the strike price and T is the maturity. And we denote the market price of European call options by $C^{(m)}(K, T)$. The error of the model price from the market price is $\widehat{C^*}(K, T) - C^{(m)}(K, T)$.

Carr and Wu [3] studied the similar problem using the FMLS (finite moment log stable) models. We disscuss the problem under the more extended model setting.

---

[2]The data used in this section were offered from Resona Bank. The author of this paper are grateful to Tetsuro Yoshimura and Kenichi Arakawa for their kindness.

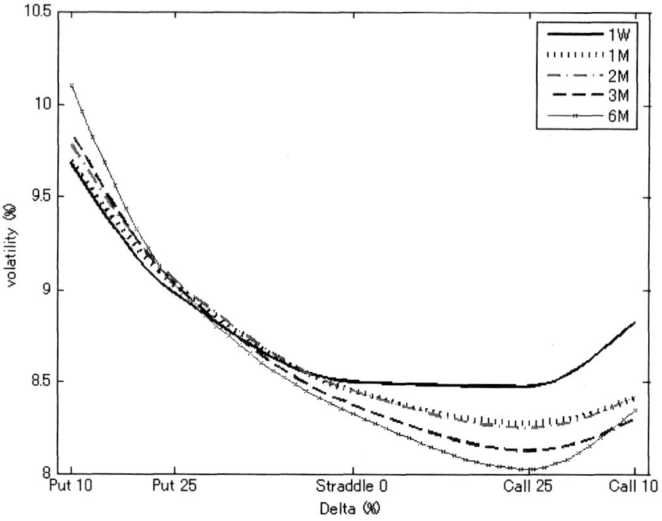

Figure 5. Volatility curve of the currency options on 05/09/2005

We adopt two kinds of measurements for the error between the model and the market option prices. The first one is the average relative percentage error (ARPE). When N theoretical prices $\{\widehat{C^*}(K_i, T_i), i = 1, 2, \ldots, N\}$ and N data $\{C^{(m)}(K_i, T_i), i = 1, 2, \ldots, N\}$ are given, then the ARPE is given by

$$\text{ARPE}^* = \frac{1}{N}\sum_{i=1}^{N} \frac{\left|\widehat{C^*}(K_i, T_i) - C^{(m)}(K_i, T_i)\right|}{C^{(m)}(K_i, T_i)} \tag{4.1}$$

The second one is the root mean-square error (RMSE) given by

$$\text{RMSE}^* = \sqrt{\frac{1}{N}\sum_{i=1}^{N} \left|\widehat{C^*}(K_i, T_i) - C^{(m)}(K_i, T_i)\right|^2} \tag{4.2}$$

4.2 Minimization Problem

Suppose that we have selected a class of [GLP & MEMM] Model (for example geometric stable process model, etc.) Then the pricing error ARPE(average relative percentage error) of the model is

$$\text{ARPE}_\alpha = \frac{1}{N}\sum_{i=1}^{N}\frac{\left|C_\alpha(K_i,T_i) - C^{(m)}(K_i,T_i)\right|}{C^{(m)}(K_i,T_i)} \tag{4.3}$$

where α is the parameter of the model, and $C_\alpha(K_i, T_i)$ is the theoretical price of it. The problem is to solve the following minimization problem.

$$\min_\alpha \text{ARPE}_\alpha = \min_\alpha \frac{1}{N}\sum_{i=1}^{N}\frac{\left|C_\alpha(K_i,T_i) - C^{(m)}(K_i,T_i)\right|}{C^{(m)}(K_i,T_i)} \tag{4.4}$$

We assume that this minimization problem has a solution, and we denote the minimum value by $\text{ARPE}^{(cal)}$. ($^{(cal)}$ means "calibration".)

We can also adopt the root mean-square error (RMSE) as another candidate for the pricing error of the model.

$$\text{RMSE}_\alpha = \sqrt{\frac{1}{N}\sum_{i=1}^{N}\left|C_\alpha(K_i,T_i) - C^{(m)}(K_i,T_i)\right|^2} \tag{4.5}$$

In this case the minimum value is denoted by $\text{RMSE}^{(\text{cal})}$.

4.3 Results

We have obtained the following results.

4.3.1 In-Sample Analysis

The time interval is 05/09/2005–04/09/2006.

The parameters of the model are estimated under the following conditions:

$$\text{SSE}(\Theta_t) := \sum_{i=1}^{n(t)} |p_t(K_i,T_i) - \eta(\Theta_t, K_i, T_i)|^2 \to \min \tag{4.6}$$

where $p_t(K_i, T_i)$ is the price of the OTC(=Over-The-Counter) call option, $\eta(\Theta_t, K_i, T_i)$ is the theoretical price of the call option based on the model, and $n(t)$ is the number of the call options at the time t. (In the case of our data, $n(t) = 25$.)

The model errors, ARPE and RMSEF

$$\text{ARPE} := \frac{1}{\sum_{t=1}^{N} n(t)}\left(\sum_{t=1}^{N}\sum_{i=1}^{n(t)}\frac{|p_{ti} - \eta_i(\Theta_t)|}{p_{ti}}\right),$$

$$\text{RMSE} := \sqrt{\frac{1}{\sum_{t=1}^{N} n(t)}\sum_{t=1}^{N}\sum_{i=1}^{n(t)}|p_{ti} - \eta_i(\Theta_t)|^2}.$$

From Table 1, we can see that the [GSP & MEMM] model is the most fitting to the data.

Here we give the table of the times of the computation with diferent models of option prices.

Table 1. In-sample Performance (sample period : 5/9/2005–4/9/2006)

Model	Measure	Calibration Error	
		ARPE	RMSE
one parameter models			
Black-Scholes Model		0.0595	0.0780
two parameter models			
FMLS		0.0549	0.0435
three parameter models			
scaled t-distribution	ESMM	0.0764	0.0558
four parameter models			
VG Process	ESMM	0.0534	0.0448
NIG Process	ESMM	0.0505	0.0544
Merton Model		0.0642	0.0405
Stable Process	MEMM	0.0392	0.0398

Table 2. Times of the computation with diferent models of option prices

Model	Measure	times(sec)
FMLS		0.072
scaled t-distribution	ESMM	0.128
Variance Gamma Process	ESMM	0.086
Normal Inverse Gaussian Process	ESMM	0.078
Merton Model		0.003
Stable Process	MEMM	1.373

4.3.2 Out-of-Sample Analysis

The time interval is 05/09/2005–04/09/2006.

The model parameters are estimated everyday as before

$$\mathrm{SSE}(\Theta_t) := \sum_{i=1}^{n(t)} |p_t(K_i, T_i) - \eta(\Theta_t, K_i, T_i)|^2 \to \min$$

The option prices of the next day are predicted using the model estimated above.

From Table 3 we can see that the [GSP & MEMM] model is the most fitting to the data.

4.3.3 Volatility-Based Calibration

In the previous sections we carried out the option price based estimation for the models. Here we will do the volatility based estimation for the models.

Table 3. Out-sample analysis

Model	Measure	Pricing Error ARPE	RMSE
one parameter models			
Black-Scholes		0.0628	0.0798
two parameter models			
FMLS		0.0575	0.0467
three parameter models			
scaled t-distribution	ESMM	0.0779	0.0584
four parameter models			
Variance Gamma Process	ESMM	0.0562	0.0480
Normal Inverse Gaussian Process	ESMM	0.0540	0.0570
Merton Model		0.0661	0.0440
Stable Process	MEMM	0.0435	0.0432

Estimation of the parameters of the model:

$$\text{SSE}(\Theta_t) := \sum_{i=1}^{n(t)} |\sigma(K_i, T_i) - \text{Vol}(\eta(\Theta_t, K_i, T_i), K_i, T_i)|^2 \to \min \tag{4.7}$$

where $\sigma(K_i, T_i)$ is the implied volatility of the call option price in the market, and $\text{Vol}(\eta, K_i, T_i)$ is the implied volatility of the given call option price η.

Table 4. Volatility-based Calibration

Model	Measure	In-sample		Out-of-sample	
		ARPE	RMSE	ARPE	RMSE
FMLS		0.0544	0.0648	0.0596	0.0678
VG Process	ESMM	0.0513	0.0599	0.0561	0.0632
Merton Model	MCMM	0.0527	0.0612	0.0580	0.0645
Stable Process	MEMM	0.0349	0.0485	0.0428	0.0525

From the above table we can see that the [GSP & MEMM] model is the most fitting to the data.

5. Conclusions and Remarks

The geometric stable process (GSP) has been supposed to be a very possible candidate for the option pricing model fitting to the strong fat tail property. We have seen that, introducing the minimal entropy martingale measure (MEMM), we could have constracted the natural pricing model [GLP & MEMM] model. And we have also seen that this model has many good properties. In fact it has the reproducibility of the volatility smile/smirk property and it is very well fitting to currency options.

Acknowledgement

The authors thank the anonymous referee for his valuable comments, and Tetsuro Yoshimura and Kenichi Arakawa for their kind support.

References

1. Barndorff-Nielsen, O. E., Normal inverse Gaussian distributions and the modeling of stock returns, Research Report No 300, Department of Theoretical Statistics, Aarhus University, 1995.
2. Carr, P., Geman, H., Madan, D. B., and Yor, M., The Fine Structure of Asset Returns: An Empirical Investigation. *Journal of Business*, **75**, (2002), 305–332.
3. Carr, P. and Wu, L., The Finite Moment Log Stable Process asd Option Pricing, *The Journal of Finance*, **58**(2), (2003), 753–777.
4. Carr, P. and Wu, L., Stochastic Skew for Currency Options, *J. of Financial Economics*, **86**(1), (2007), 213–247.
5. Cont, R and Fonseca, J., Dynamics of Implied Volatility Surfaces, *Quantitative finance*, **2**, (2002), 45–60.
6. Edelman, D., A Note: Natural Generalization of Black-Scholes in the Presence of Skewness, Using Stable Processes, *ABACUS*, **31**(1), (1995),113–119.
7. Esche, F. and Schweizer, M., Minimal entropy preserves Levy property:How and why, *Stochastic Processes and their Applications*, **115**, (2005), 299–327.
8. Fama, E. F., Mandelbrot and the Stable Paretian Hypothesis. *J. of Business*, **36**, (1963), 420–429.
9. Frittelli, M., The Minimal Entropy Martingale Measures and the Valuation Problem in Incomplete Markets, *Mathematical Finance*, **10**, (2000), 39–52.
10. Fujiwara, T. and Miyahara, Y., The Minimal Entropy Martingale Measures for Geometric Lévy Processes, *Finance and Stochastics*, **7**, (2003), 509–531.
11. Hubalek, F. and Sgarra, C., Esscher transforms and the minimal entropy martingale measure for exponential Lévy models, *Quantitative Finance*, **6**, (2006), 125–145.
12. Madan, D., Carr, P. and Chang, E., The variance gamma process and option pricing. *European Finance Review*, **2**, (1998), 79–105.
13. Madan, D. and Seneta, E., The variance gamma (vg) model for share market returns. *Journal of Business*, **63**(4), (1990), 511–524.
14. Merton, R. C., Option Pricing when Underlying stock returns are discontinuous, *Journal of Financial Economics*, **3**, (1976), 125–144.
15. Miyahara, Y., Canonical Martingale Measures of Incomplete Assets Markets, in *Probability Theory and Mathematical Statistics: Proceedings of the Seventh Japan-Russia Symposium, Tokyo 1995 (eds. S. Watanabe et al.),* (1996), 343–352.
16. Miyahara, Y., Minimal Entropy Martingale Measures of Jump Type Price Processes in Incomplete Assets Markets. *Asian-Pacific Financial Markets*, **6**(2), (1999), 97–113.
17. Miyahara, Y., [Geometric Lévy Process & MEMM] Pricing Model and Related Estimation Problems, *Asia-Pacific Financial Markets*, 8(1), (2001), 45–60.
18. Miyahara, Y., The [GLP & MEMM] Pricing Model and Related Problems, *Proceedings of the 5th Ritsumeikan International Symposium ' Stochastic Processes and Applications to Mathematical Finance'* (eds. J. Akahori et al), (2006), 125–156.
19. Miyahara, Y. and Moriwaki, N., Volatility smile/smirk properties of [GLP & MEMM]

models, *RIMS Kokyuroku 1462*, (2006), 156–170.
20. Moriwaki, N. and Miyahara, Y., Empirical Performance of the [GSP & MEMM] Option Pricing Models, *MTEC Journal*, **20**, (2008), 71–88. (In Japanese)
21. Rachev, S. and Mittnik, S., *Stable Paretian Models in Finance*, Wiley, 2000.

The Impact of Momentum Trading on the Market Price and Trades

Katsumasa Nishide

Interdisciplinary Research Center, Yokohama National University
E-mail: knishide@ynu.ac.jp

This paper theoretically analyzes a financial market with momentum trading in a market microstructure framework, and investigates how momentum trading affects the market price and trades. When momentum traders dominate the market, private information owned by rational traders may not be incorporated into the price, and a market bubble may occur. It is shown that the reliability of a firm's public announcement such as an earnings forecast is a key factor to avert the market bubble.
Key words: Market microstructure, momentum trading, market bubble.

1. Introduction

Milgrom and Stokey [18] proved that if all traders in the market are rational and fully exploit their information available to them, i.e., if there is no noise trading in the market, no trade occurs after the initial trades. This happens even when traders receive different signals on the fundamental value of the asset. The *No Trade Theorem* holds because the revealed price is the sufficient statistic of the fundamental value, and there can be no trade motivation under the Pareto optimum allocation. Conversely, the theorem implies that in actual financial markets, stocks and bonds are actively traded over periods thanks to so-called noise traders, who trade for exogenous or irrational reasons that cannot be explained by the traditional economic theory. Black [4] emphasized the role of noise traders by saying that "Noise makes trading in financial markets possible, and thus allows us to observe prices for financial assets." Thus, it is of great importance for financial researchers and practitioners to analyze the market with irrational traders.

In the market microstructure theory, many kinds of traders have been investigated. In many papers, irrational trading is simply modeled as a normal distribution with mean zero, independent of all other random variables (e.g., Kyle [17], Admati [1]). Under this assumption, the price becomes an imperfect signal on the true value of the asset (hereafter we call it the liquidation value), and noise traders merely stand at the opposite trading position to rational traders so that the market clears all orders. Noise trading in these studies only plays a passive role in the market.

In actual markets, however, there are other kinds of traders whose trading strategies seem contradictory to the standard theory of finance, but have a certain impact on the market. In behavioral finance, many papers tried to model a different type of investors and explain anomalies observed in the financial market. For example, Daniel *et al.* [8] analyzed the effect of the overconfidence and self-attribution of investors, and showed that overconfidence causes negative long-lag autocorrelations, while self-attribution adds positive short-lag autocorrelations. See Barberis and Thaler [3] or Subrahmanyam [19] for an extensive survey of the literature.

In this paper, we study the effect of *momentum trading* on the market price. We here define the momentum trading to be the strategy that trades according to the past price process. Typical examples are "trend-chasing strategy" and "contrarian strategy." Even professional traders such as technical analysts follow these strategies, which can cause a big price impact. The aim of this paper is not to examine how and why there are momentum traders in the market and they survive, but to theoretically investigate how the momentum trading affects the market price, informational efficiency, and so on. We also try to show how to decrease the effect of momentum trading, since momentum trading may cause a kind of instability in the market.

The theoretical study of momentum trading is not new in the literature. For example, De Long *et al.* [9] studied a market with momentum trading. Their results are very similar to ours, and this paper can be thought of as their extension in the following sense. In De Long *et al.* [9], informed traders privately observe a noisy signal on the liquidation value. They assume that noise term in the signal is a discrete random variable, while the liquidation value of the asset follows a continuous random variable. We model the liquidation value and all noise terms to be continuous random variables, more specifically (independent) normals. Thanks to the assumption that all random variables follow normals, we can discuss how to avoid a market bubble as in Section 4. In De Long *et al.* [9], how a market bubble occurs in their model is focused on, its countermeasures are not considered. This is one of advantages in this paper.

Hong and Stein [12] also analyzed a market with momentum trading, and showed that the prices underreact in the short run and overreact in the long run, as suggested in empirical papers. In their model, however, newswatchers, who can observe private signals on the liquidation value as rational traders in this paper, only maximize their one-period profit although they trade at multiple periods. Due to the assumption of their myopic behavior, newswatchers lose money by momentum traders in some parameter setting. But it is too restrictive to suppose that traders with superior information trade in order to maximize locally. Moreover, they only analyze the stationary price dynamics. Therefore, bubble phenomena due to momentum trading cannot be studied in their setting.

In this paper, we extend Kim and Verrecchia [16], and construct a market model where a risky asset is traded in 2 periods. In our model, rational traders, who observe a private signal on the liquidation value of the asset, maximize their conditional expected utility given all available information. Each rational trader has an exponential utility function with a constant risk tolerance coefficient (reciprocal of absolute risk-aversion coefficient). This kind of setting is called a CARA-Normal model, and popular in the market microstructure literature for its tractability.

Contrary to Kim and Verrecchia [16], we introduce momentum traders into our model in addition to rational traders. Momentum traders are assumed to trade an asset according to the past price path. More specifically, we assume that the aggregated order amount from momentum traders at period 2, say q_2, is given by

$$q_2 = \phi(P_1 - P_0),$$

where P_t denotes the market price at period t. When $\phi > 0$, the price increase at the previous period leads to buy orders by momentum traders.[1] Therefore, a positive ϕ assumes the situation in which so-called trend-chasing traders actively trade in the market. Similarly, a negative ϕ means that there are a significantly many contrarian traders who submit sell orders when the price has increased participate in the market.[2]

Our main results are the followings. First, momentum trading affects the market price only through the ratio ϕ/r, where r is the average of the risk tolerance coefficient of all rational traders. Intuitively, the coefficient r represents how aggressively rational traders trade an asset. When r is large, rational traders trade actively based on their private information, and the effect of momentum traders on the market price is diminishing. On the other hand, when the magnitude of ϕ is large in comparison with the risk tolerance of rational traders, the impact of momentum trading becomes large, and the market can be unstable and informationally inefficient.

Second, if ϕ/r is large enough, the price at the earlier period behaves in a counterintuitive way. For example, when ϕ/r is large, the price becomes lower as the liquidation value takes a higher value. In a usual situation, the market price should be increasing as the fundamental value of the asset is larger. This is because the market price incorporates some private information owned by rational

---

[1] We follow the convention that a positive order amount means a net buy order, and vice versa.

[2] Hong and Stein [12] consider the momentum traders' optimization problem with respect to ϕ. Note again that the aim of this paper is to study the effect of momentum trading on the market price, not how and why momentum traders are in the market. Therefore, we set the value of ϕ exogenously and analyze how ϕ affects the market price.

traders through their trades. We find that this is not the case when there are many trend-chasing traders in the market.

Third, as ϕ diverges to positive or negative infinity, the price at period 1 converges to the price at period 0, the unconditional expectations. This means that, when momentum traders dominate the market, no matter whether they are trend-chasing or contrarian, the market price has little information of the private signals, i.e. it is informationally inefficient.

Fourth, the price at the earlier period is dependent on the precision of a public signal at the later period. More precisely, if the precision of public signal at period 2 is high, then rational traders trade more aggressively based on their private signals at period 1. As a result, the market price at period 1 becomes stable and informationally efficient, even though rational traders do not observe the realized public signal. This observation indicates that if market participants believe that pubic information such as a future earnings forecast is precise enough, the present price reflects enough private information available to rational traders. This finding has an important economic implication. If financial authorities consider the efficiency and stability of the market, one of effective measures is to assure the transparency and reliability of accounting information or other IR activities. If the public information in the future is reliable, rational traders trade actively based on their private signals even before they receive the actual public announcement, and so the private information is properly incorporated into the market price.

These findings indicates that the introduction of momentum traders into the model of Kim and Verrecchia [16] significantly changes the properties of the market prices and trades. Also, the results obtained in this paper give new insights into empirical findings in the previous literature. There are many studies that reported the price momentum in stock returns, especially in short-run returns. Some papers attribute the price momentum into slow diffusion of news (e.g., Jegadeesh and Titman [14], Chan *et al.* [6], Hong *et al.* [13]), and others to psychological tendency of market participants (e.g., Kausar and Taffler [15], Doukas and Petmezas [10]). Recently, Chan [7] provided the evidence that the market underreacts to firm-specific news released publicly, and overreacts to news implied by price changes, and argued that this phenomenon is partly due to the psychological bias as in Daniel *et al.* [8]. Zhang [20] investigated the role of information uncertainty, observed that information uncertainty yields price continuation anomalies, and concluded that the incompleteness of the initial market reaction to new public news is the main reason for the price momentum, and that the uncertainty delays the flow of information into stock prices. Our model also obtains similar results to theirs, but has another implication that the price momentum is caused not only by momentum traders but also by rational traders through their strategic behavior as follows.

If rational investors know that there are momentum traders in the market, they take advantage of the existence of momentum traders, and follow the strategy that can make a profit from momentum traders. They do not utilize their information on the liquidation value as in the case that there are only rational traders in the market. When rational investors adopt such policy, the market does not incorporate the information owned by rational traders into the price, meaning less informative price. Many papers in behavioral finance attribute the price anomalies to irrational or bounded-rational traders whose strategy cannot be explained by the traditional finance theory. However, our observation suggests that the price anomalies reported in Chan [7] and Zhang [20] may be influenced by not only momentum traders but also rational investors through their strategies.

This paper is organized as follows. Section 2 describes the model, and Section 3 provides the main results. In Section 4, we discuss in detail about our results and their implications with numerical examples. Section 5 provides the conclusion of this paper. The detailed proofs of the theorem and propositions are given in Appendix.

2. Model Setup

Our market model is based on Kim and Verrecchia [16]. We consider a securities market where one risky asset is traded. Risk-free rate is assumed to be zero for simplicity. Trades of an risky asset take place at $t = 1$ and 2. After period 2, the liquidation value is realized, all positions are cleared, and consumptions occur. The liquidation value, denoted by v, is normally distributed with mean P_0 and precision (reciprocal of the variance) ρ_{v_0}. The distribution of the liquidation value is known to all market participants. The price of the asset at period t is denoted by P_t.[3] In the market, there are two types of traders: rational traders and momentum traders.

We assume that there is a continuum mass of rational traders indexed by $j \in [0, 1]$. Each trader behaves as a price-taker. We denote the position of rational trader j by x_{jt}. The initial position x_{j0} is exogenously and randomly given to rational trader j. The aggregated endowment of the risky asset, denoted by

$$x = \int_{j\in[0,1]} x_{j0}\mathrm{d}j,$$

is not known to individual traders, and all traders believe that x is normally distributed with mean 0 and precision ρ_x, and independent of all random variables. The randomness of the risky asset supply captures the fact that securities markets are generally subject to random demand and supply fluctuations due to changing liquidity needs, weather, political situations, etc.

---

[3]We implicitly assume that the unconditional mean P_0 is the price before period 1, and that the price P_0 reflects all available information at period 0. This assumption is also adopted by De Long *et al.* [9]

Before period 1, trader j observes his/her private signal on the liquidation value, say s_j, before the market opens. The signal s_j is a noisy unbiased signal of the form

$$s_j = v + \epsilon_j,$$

where the noise term ϵ_j is normally distributed with mean 0 and precision ρ_{ϵ_j}, and independent of v.

In addition to the private signal, all traders observe a public announcement about the liquidation value at each period. We assume that the announcement at each period is given by

$$y_t = v + \eta_t, \quad t = 1, 2,$$

where η_t is normally distributed with mean 0 and precision ρ_{η_t}, and independent of all other random variables. The public signal can be thought of as a kind of earnings forecasts released by the firm. After the announcement of the public signal, rational traders update their conditional distribution of the liquidation value v, and submit an order based on their updated belief.

Rational traders trade a risky asset to maximize their conditional expected utilities, a negative exponential function with risk tolerance coefficient r_j. More formally, the problem of rational trader j is given by

$$\max \mathbb{E}\left[-e^{-\frac{1}{r_j}W_j} \middle| \mathcal{I}_{jt}\right], \quad t = 1, 2,$$

where $W_j = \sum_{t=0}^{2}(P_{t+1} - P_t)x_{jt}$, and $\mathcal{I}_{jt}$ is all available information of rational trader j at period t. Note that each rational trader can utilize the observed value of P_t as a signal on the liquidation value v when they trade at period t.[4]

Momentum traders trade at period 2 according to the price movement from period 0 to 1. The aggregated amount of the orders from momentum traders, denoted by q_2, is assumed to be

$$q_2 = \phi(P_1 - P_0).$$

The parameter ϕ represents how much momentum traders respond to the past price movement. If $\phi > 0$, we assume that there are many trend-chasing traders in the market, while if $\phi < 0$, the order from contrarian traders dominates the momentum

[4]Many extant papers in market microstructure assume a so-called CARA-normal setting, meaning that all random variables are normal and all agents have exponential utility functions. In this setting, the liquidation value and the final wealth can be negative. However, when the expected values are sufficiently large, the results have no quantitative difference, and so we follow the setting adopted by existing papers.

trading. The modeling of momentum trading is the same as Hong and Stein [12], except that ϕ is exogenously given in this paper.

Finally, we define the market price in this model.

Definition 2.1. The market price satisfies the following conditions:

(i) Each rational trader maximizes his/her conditional expected utility given all available information.

(ii) The amount of total demand is equal to the supply of the asset:

$$x = \begin{cases} x_1 & \text{for } t = 1 \\ x_2 + q_2 & \text{for } t = 2, \end{cases} \tag{1}$$

where $x_t := \int_{j\in[0,1]} x_{jt} \mathrm{d}j$.

Figure 1 concisely illustrates the time line of the model.

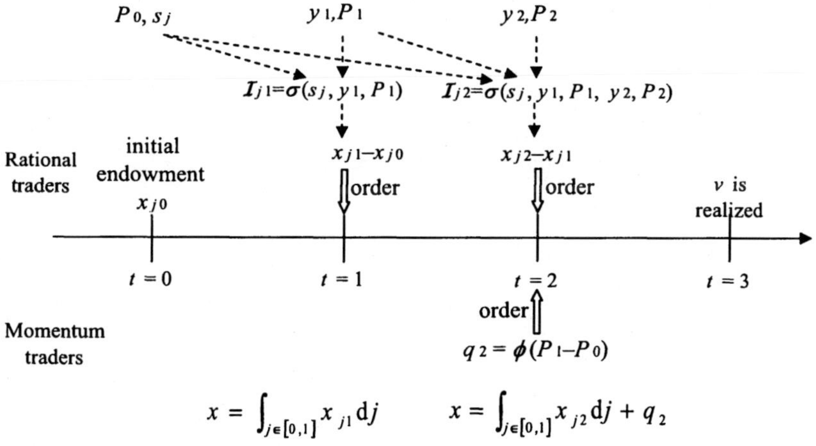

Figure 1. Time line of the model. The dotted arrows indicate the available information to rational traders. Note that the price at each period has some signal on the liquidation value, and that the initial position x_{j0} has no information for the liquidation value.

Before the market opens, each rational trader observes P_0, the ex-ante expectation of the liquidation value as a public signal, and the noisy private signal s_j that is not observable by other traders. At period 1, each trader observes y_1, the public information on the liquidation value and P_1, the market price. The orders are cleared and settled only after the total order equates the supply of the asset.

Similar events happen at period 2, except that there are momentum traders in the market. At period 3, the liquidation value is realized, and consumptions occur.

Note again that the price at each period has some signal on the liquidation value. This is because through orders from rational traders, the private signals owned by rational traders are partially incorporated through the price formation process that depends on the order balance in the market.

3. Market Prices

In this section, we derive the market price formulae, and analyze the properties of the market. The proof of the following theorem is given in Appendix A.

Theorem 3.1. *Define*

$$r := \int_j r_j \mathrm{d}j, \quad \rho_\epsilon := \frac{1}{r}\int_j r_j \rho_{\epsilon_j} \mathrm{d}j,$$

$$\rho_{v_1} := \rho_{v_0} + \rho_{\eta_1} + \rho_\epsilon + (r\rho_\epsilon)^2 \rho_x, \quad \rho_{v_2} := \rho_{v_1} + \rho_{\eta_2}$$

and

$$\kappa := \frac{\phi}{r\rho_{v_2}}.$$

(i) *Suppose* $\kappa \neq 1$. *Then, the market price at period 1 is given by*

$$P_1 = \left(\frac{\rho_{v_0}}{(1-\kappa)\rho_{v_1}} - \frac{\kappa}{1-\kappa}\right) P_0 + \frac{\rho_{\eta_1}}{(1-\kappa)\rho_{v_1}} y_1 + \frac{\rho_\epsilon + r^2\rho_\epsilon^2\rho_x}{(1-\kappa)\rho_{v_1}} v - \frac{1 + r^2\rho_\epsilon\rho_x}{(1-\kappa) r\rho_{v_1}} x, \tag{2}$$

and the market price at period 2 is given by

$$\begin{aligned} P_2 = {} & \left[\left(\frac{1}{\rho_{v_2}} + \frac{\kappa}{1-\kappa}\frac{1}{\rho_{v_1}}\right)\rho_{v_0} - \frac{\kappa}{1-\kappa}\right] P_0 \\ & + \left(\frac{1}{\rho_{v_2}} + \frac{\kappa}{1-\kappa}\frac{1}{\rho_{v_1}}\right)(\rho_\epsilon + r^2\rho_\epsilon^2\rho_x) v \\ & + \left(\frac{1}{\rho_{v_2}} + \frac{\kappa}{1-\kappa}\frac{1}{\rho_{v_1}}\right)\rho_{\eta_1} y_1 + \frac{\rho_{\eta_2}}{\rho_{v_2}} y_2 \\ & - \left(\frac{1}{\rho_{v_2}} + \frac{\kappa}{1-\kappa}\frac{1}{\rho_{v_1}}\right)\frac{1 + r^2\rho_\epsilon\rho_x}{r} x. \end{aligned} \tag{3}$$

When $\kappa = 1$, *the market cannot clear the orders.*

(ii) Let

$$\theta := \frac{(1-\kappa)\rho_{v_1}}{\rho_\epsilon(1+r^2\rho_\epsilon\rho_x)}P_1 - \frac{\rho_{v_0}-\kappa\rho_{v_1}}{\rho_\epsilon(1+r^2\rho_\epsilon\rho_x)}P_0 - \frac{\rho_{\eta_1}}{\rho_\epsilon(1+r^2\rho_\epsilon\rho_x)}y_1.$$

The optimal demands of rational trader j at periods 1 and 2 are given by

$$\begin{aligned} x_{j1}^* = r_j \Bigg[& \frac{\rho_{v_2}}{\rho_{\eta_2}}(\rho_{v_0}-\kappa\rho_{v_1})P_0 + \rho_{\epsilon_j}s_j \\ & + \left(\frac{\rho_{v_2}}{\rho_{\eta_2}}(\rho_\epsilon + r^2\rho_\epsilon^2\rho_x) - \rho_\epsilon\right)\theta \\ & - \left(\frac{\rho_{v_2}}{\rho_{\eta_2}}(1-\kappa)\rho_{v_1} - (\rho_{\epsilon_j}-\rho_\epsilon)\right)P_1 \Bigg] \end{aligned} \tag{4}$$

and

$$x_{j2}^* = r_j\left[\rho_{v_0}P_0 + \rho_{\epsilon_j}s_j + \rho_{\eta_1}y_1 + \rho_{\eta_2}y_2 + r^2\rho_\epsilon\rho_x\theta - \rho_{j2}P_2\right], \tag{5}$$

respectively. Here, $\rho_{j2} := \rho_{v_0} + \rho_{\epsilon_j} + \rho_{\eta_1} + \rho_{\eta_2} + r^2\rho_\epsilon\rho_x$.

Note from (2) and (3) that momentum trades only affect the market prices through the term $\kappa := \phi/r\rho_{v_2}$. It is verified that (3) can be rewritten as

$$\begin{aligned} P_2 = & \frac{\rho_{v_0}}{\rho_{v_2}}P_0 + \frac{\rho_{\eta_1}}{\rho_{v_2}}y_1 + \frac{\rho_{\eta_2}}{\rho_{v_2}}y_2 + \frac{\rho_\epsilon(1+r^2\rho_\epsilon\rho_x)}{\rho_{v_2}}v \\ & - \frac{1+r^2\rho_\epsilon\rho_x}{r\rho_{v_2}}x + \kappa(P_1 - P_0). \end{aligned} \tag{6}$$

The last term of (6) indicates that κ describes the sensitivity of P_2 with respect to the price movement $P_1 - P_0$. When ϕ is positive, κ is also positive, and the price increase in the previous period induces the upward movement of P_2. Therefore, κ is regarded as a parameter that connects the prices of the two consecutive periods through momentum trading.

When $\kappa = 1$, the market price cannot be formed. To see this, the aggregated demand of all rational traders is given by

$$\begin{aligned} x_1^* = r\Bigg[& \frac{\rho_{v_0}\rho_{v_2}(1-\kappa\rho_{v_1})}{\rho_{\eta_2}}P_0 + \frac{\rho_{\eta_1}\rho_{v_2}}{\rho_{\eta_2}}y_1 \\ & + \frac{\rho_{v_1}\rho_\epsilon + \rho_{v_2}\rho_\theta}{\rho_{\eta_2}}\theta + \rho_\epsilon v - \frac{\rho_{v_1}}{r\rho_{\eta_2}}x - \frac{\rho_{v_1}\rho_{v_2}(1-\kappa)}{\rho_{\eta_2}}P_1\Bigg]. \end{aligned} \tag{7}$$

where $\rho_\theta := r^2\rho_\epsilon^2\rho_x$. From the last term in (7), we observe that the demand of rational traders is independent of P_1 if $\kappa = 1$. Therefore, the market cannot clear the orders.

Next, we take a closer look at the proprieties of the market price, the trading volume, and so on.

Proposition 3.1.

(i) The price change at period 2 is expressed as

$$P_2 - P_1 = \frac{\rho_{\eta_2}}{\rho_{v_2}}\left[y_2 - \int_{j\in[0,1]} \mathbb{E}[v|\mathcal{I}_{j1}]\mathrm{d}j + \frac{x}{r\rho_{v_1}}\right]. \tag{8}$$

(ii) The trading amount of trader j at period 2 is expressed as

$$x_{j2} - x_{j1} = -r_j(\rho_{\epsilon_j} - \rho_\epsilon)(P_2 - P_1) - \frac{r_j}{r}q_2 \tag{9}$$

The first part of Proposition 3.1 is exactly the same as the result obtained in Kim and Verrecchia [16]. The price reaction to a public announcement is proportional to the importance of the announced information relative to the average posterior beliefs of rational traders and the surprise contained in the announced information plus noise. That is,

$$P_2 - P_1 = \text{Surprise+Noise},$$

where

$$\text{Surprise} := \frac{\rho_{\eta_2}}{\rho_{v_2}}\left(y_2 - \int_{j\in[0,1]} \mathbb{E}[v|\mathcal{I}_{j1}]\mathrm{d}j\right)$$

and

$$\text{Noise} := \frac{\rho_{\eta_2}}{\rho_{v_2}}\frac{x}{r\rho_{v_1}}.$$

Note that the price movement does not include the effect of momentum trading. This finding indicates that the price at period 1 fully reflects the impact of momentum trading, although momentum trading only appears in period 2. Even if market participants believe that there is momentum trading only in the future, the price effectively contains the impact before momentum traders arrive at the market.

The second statement in Proposition 3.1 says that the trading volume at period 2 is decomposed into three parts: surprise term, noise effect, and momentum term. That is

$$x_{j2} - x_{j1} = -(\text{Surprise+Noise})\text{+Momentum},$$

where

$$\text{Momentum} := \frac{r_j}{r}q_2.$$

This decomposition implies that there are two kinds of motivations for each rational trader to trade the asset. The first one is the information effect appeared in the first term of (9). Each rational trader updates their belief based on the information available to him/her. If ρ_{ϵ_j} is smaller than ρ_ϵ and the price moves upward, he/she updates their belief and buys the asset, since the private information of rational trader j is not accurate in comparison with those of other traders. If $\rho_{\epsilon_j} > \rho_\epsilon$, trader j will think that his/her own private signal is reliable, and that the price process does not contain much information on the liquidation value. In this case, trader j buys when the price moves downward, and vice versa.

The second term in (9) is regarded as the momentum effect. In this setting, rational traders know the existence of momentum traders in advance. Hence, rational traders take the effect of momentum trading at period 2 into account when they trade. For example, when $\phi > 0$ and $P_1 > P_0$, rational traders know at the end of period 1 that the demand-supply balance at period 2 becomes tight because of the buy orders from momentum traders. Then, the price is likely to go up regardless of the liquidation value. Therefore, rational traders can earn a profit by buying at period 1 and selling at period 2. The last term of (9) explains this effect.

Next, we analyze the effect of momentum trading on the prices. The following proposition can be obtained by simply taking partial derivatives.

Proposition 3.2. *When $\phi > r\rho_{v_2}$, we have*

$$\frac{\partial P_t}{\partial v} < 0, \quad \frac{\partial P_t}{\partial y_1} < 0, \quad \frac{\partial P_t}{\partial x} > 0. \tag{10}$$

for $t = 1, 2$.

Proposition 3.2 implies that if ϕ is large enough, the prices at both periods behave in a counterintuitive way. For example, the first inequality of (10) indicates that the prices at both periods become lower as a higher value of v is realized. In Kyle [17], the private information owned by the informed trader is gradually incorporated through their orders. Thus, when each informed trader observes a high s_j for example, the market price at each period should be higher in a usual situation. The first inequality contradicts a natural intuition that the market price partially reflects private information owned by market participants. The reason of this price behavior is that if trend-chasing traders dominate the market, then rational traders know that momentum traders sell at period 2 when P_1 falls, and they do not trade based on their private information, but try to earn profits at the expense of momentum traders instead.

The second inequality of (10) says that when market participants believe that the liquidation value is expected to be higher, the market price becomes lower with momentum trading. This means that in a market with trend-chasing momentum traders, even though a firm releases an announcement that its business is doing well, the price decrease in spite of the good news. In most of noisy rational expectations models with multiple periods, the price gradually converges from the initial price to the liquidation value as in Kyle [17]. In a market with many trend-chasers, the public signal in an intermediate period affects the price in an opposite way.

The third inequality of (10) is also contrary to the ordinary situation. A high value of x means a large supply of the risky asset. Then, the balance of demand-supply becomes loose, and the market price should fall in a normal setting. When trend-chasers actively trade in the market, a large supply of the asset may lead to a higher market price.

Next, we investigate the impact of momentum trading on the uncertainty of the price movement, and the serial correlation.

Proposition 3.3. *The variances of the price movement are given by*

$$\mathbb{V}[P_1 - P_0] = \frac{\frac{[\rho_{\eta_1} + \rho_\epsilon(1+r^2\rho_\epsilon\rho_x)]^2}{\rho_{v_0}} + \rho_{\eta_1} + \frac{[1+r^2\rho_\epsilon\rho_x]^2}{r^2\rho_x}}{[(1-\kappa)\rho_{v_1}]^2}$$

and

$$\mathbb{V}[P_2 - P_1] = \left(\frac{\rho_{\eta_2}}{\rho_{v_1}\rho_{v_2}}\right)^2 \left[\rho_{v_0} + \rho_{\eta_1} + \left(\frac{1 + r^2\rho_\epsilon\rho_x}{r}\right)^2 \frac{1}{\rho_x}\right] + \frac{\rho_{\eta_2}}{\rho_{v_2}^2}.$$

In particular, when $\phi \to \pm\infty$, we have $\mathbb{V}[P_1 - P_0] \to 0$ and $P_1 \to P_0$. The autocorrelations of the price movement are given by

$$\mathbb{C}[P_2 - P_1, P_1 - P_0] = -\frac{\rho_{\eta_2}}{(1-\kappa)\rho_{v_1}^2\rho_{v_2}} \frac{1 + r^2\rho_\epsilon\rho_x}{r^2\rho_x} \tag{11}$$

and

$$\mathbb{C}[P_3 - P_2, P_2 - P_1] = -\,\mathbb{C}[P_2 - P_1, P_1 - P_0], \tag{12}$$

where $P_3 = v$.

Proposition 3.3 indicates that momentum trading affects the uncertainty of the price movement at the earlier period, not at the later period. This observation is consistent with Proposition 3.1. The price at period 1 fully reflects the impact of momentum trading, although momentum traders participate in the market only in period 2. In other words, the market price at period 1 incorporates all available information, including the structure of momentum trading at period 2. This is because all rational traders take into account the fact that momentum traders participate in the market at period 2, and fully exploit their irrational behavior at period 1.

When $\phi \to \pm\infty$, the price at period 1 converges to the unconditional expectation, which means that price has no new information on the liquidation value. When momentum trader dominates the market, the price becomes less informative. In the case of $\phi \to \pm\infty$, all rational traders trade based on the momentum effect, and do not trade according to their private signal. Equation (9), which describes the trading volume at period 2, confirms this finding.

The serial correlations in our model clearly explain the effect of the momentum trading on the price movement. The price increase at period 1 induces buy orders by momentum traders, and the price decrease at period 1 leads to their sell orders. Therefore, the correlation (11) should be positive when κ is large enough. On the other hand, information asymmetry vanishes and the market price incorporates all available information at period 3. Consequently, the effect of the momentum trading disappears at the period, and the correlation (12) should be negative.

Finally, we take a look at the informational efficiencies of the market price. We define the informational efficiency at period t to be $\mathbb{E}[(v - P_t)^2]$.[5]

Proposition 3.4. *The informational efficiency at period 1 is given by*

$$\mathbb{E}[(v - P_1)^2] = \left[\frac{\kappa\rho_{v_1}}{\rho_{v_1}(1-\kappa)}\right]^2 \frac{1}{\rho_{v_0}} + \left[\frac{1 + r^2\rho_\epsilon\rho_x}{r\rho_{v_1}(1-\kappa)}\right]^2 \frac{1}{\rho_x}$$

and that at period 2 is equal to

$$\begin{aligned}\mathbb{E}[(v - P_2)^2] &= \left(\frac{\kappa}{1-\kappa}\right)^2 \frac{1}{\rho_{v_0}} + \left(\frac{\rho_{v_0}}{\rho_{v_2}}\right)^2 P_0^2 \\ &\quad + \left(\frac{(1 + r^2\rho_\epsilon\rho_x)}{r\rho_{v_2}}\frac{\rho_{v_1} + \kappa\rho_{\eta_2}}{\rho_{v_1}(1-\kappa)}\right)^2 \frac{1}{\rho_x} \\ &\quad + \left(\frac{\rho_{\eta_1}}{\rho_{v_2}}\frac{\rho_{v_1} + \kappa\rho_{\eta_2}}{\rho_{v_1}(1-\kappa)}\right)^2 \frac{1}{\rho_{\eta_1}} + \left(\frac{\rho_{\eta_2}}{\rho_{v_2}}\right)^2 \frac{1}{\rho_{\eta_2}}.\end{aligned}$$

Proposition 3.4 says that if momentum traders dominate the market, i.e., $|\kappa|$ is large enough, informational efficiency worsens. That is, the price at period 1 contains little private information observed by each rational trader. This observation reconfirms the results obtained in Propositions 3.1 and 3.3.

Here, we summarize the results obtained in the above analysis:

[5] In a Kyle-type model, the informational efficiency is define by $\mathbb{V}[v|\mathcal{I}_t]$, where $\mathcal{I}_t$ denotes the public information at period t. If the price satisfies $P_t = \mathbb{E}[v|\mathcal{I}_t]$, then

$$\mathbb{E}[(v - P_t)^2] = \mathbb{E}[(v - \mathbb{E}[v|\mathcal{I}_t])^2] = \mathbb{E}[\mathbb{E}[(v - \mathbb{E}[v|\mathcal{I}_t])^2|\mathcal{I}_t]] = \mathbb{V}[v|\mathcal{I}_t],$$

where we have used in the last equation the fact that the conditional variance is independent of the realization of v when v follows a normal. Hence, our definition is consistent with existing studies.

- Rational traders have two trade motivations. One is based on his/her private information, and the other is on the utilization of momentum trading.
- Even if momentum traders are in the market at period 2, the market price fully reflects the impact of momentum trading at period 1.
- When momentum traders dominate in the market, the price behaves in a counterintuitive way.
- When the impact of momentum trading is large, the price at period 1 reflects less private information observed by rational traders.

4. Discussions

In this section, we present some numerical examples, and show that releasing reliable public information is an effective way to avert a market bubble.

In a market with large $\phi > 0$ and small r, a market bubble may happen. We have observed from (2) that when ϕ is large and r is small, rational traders utilize the existence of momentum traders rather than buy and sell based on the private signals. This means that the market shows a kind of price manipulation. The price is largely affected by the momentum trading, and do not reflect the private information.

Now we provide the numerical examples that effectively describe the effect of momentum trading. The common parameters and variables are presented in Table 1. Note that P_0, the initial price, is +1 and the liquidation value is −1. Moreover,

Table 1. Base case parameters

parameter	r	ρ_{v_0}	ρ_ϵ	ρ_x	ρ_{η_1}	ρ_{η_2}	P_0	$P_3 = v$	x	y_1	y_2
value	1	1	1	1	1	1	+1	−1	0	−1	−1

the public signals at both periods are equal to the true liquidation value. Hence, the price should be decreasing gradually if the price reflects the private and public information.

Example 1. In Example 1, we set $\phi = 0$, meaning that there is no momentum trading in the market. Figure 2 depicts the price path from period 0 to period 3.

We observe from Figure 2 that if there is no momentum trading, the market price is gradually decreasing over time as we have expected. This is because public and private information observed by rational traders is gradually incorporated into the price through their trades.

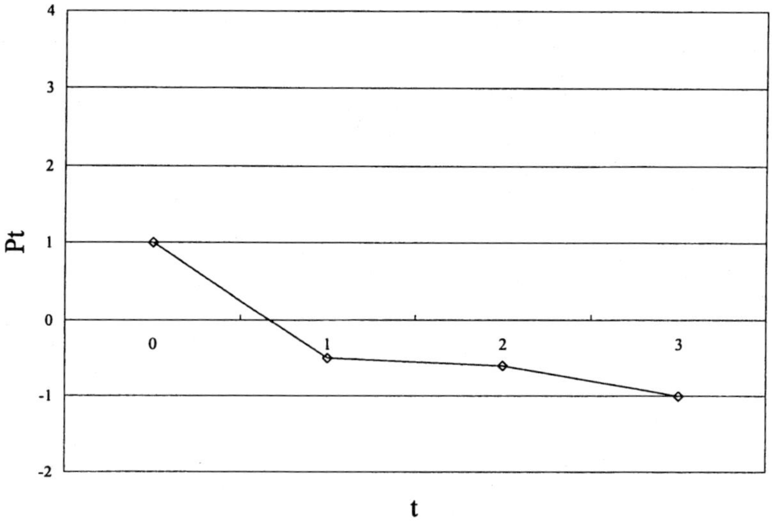

Figure 2. Price path when there is no momentum trading. In this case, the price is gradually decreasing because private and private information is incorporated into price through the trades by rational traders.

Example 2. In Example 2, we set $\phi = 8$. The setting of other parameters is the same as in Example 1. In this case, κ, representing the impact of the momentum trading on the price, is equal to 1.6. The price path in this case is presented in Figure 3.

We find from Figure 3 that the price at period 1 rises sharply, although the liquidation value is lower than the initial price. The price at period 2 still remains higher than the initial price P_0. Finally, the liquidation value reveals, and the price sharply falls, meaning a bubble burst of the market. This result indicates that if trend-chasing strategy is dominant in the market, it causes a market bubble and makes the market unstable.

The intuition is as follows. We observe from (9) that rational traders' profit opportunity lies in the two sources: the public and private signals on the liquidation value, and the existence of momentum traders' strategy. When r is small or ϕ is large, rational traders take of the behavior of momentum traders into account rather than their private signals. This means that the orders from rational traders do not reflect their private information, and the price is determined mainly based on the momentum effect. Then, the information is not incorporated into the market price, meaning that the market price is less informative. Finally, after the market is closed, the liquidation value reveals, and market participants learn that the price does not reflect the private information, and the price sharply decreases.

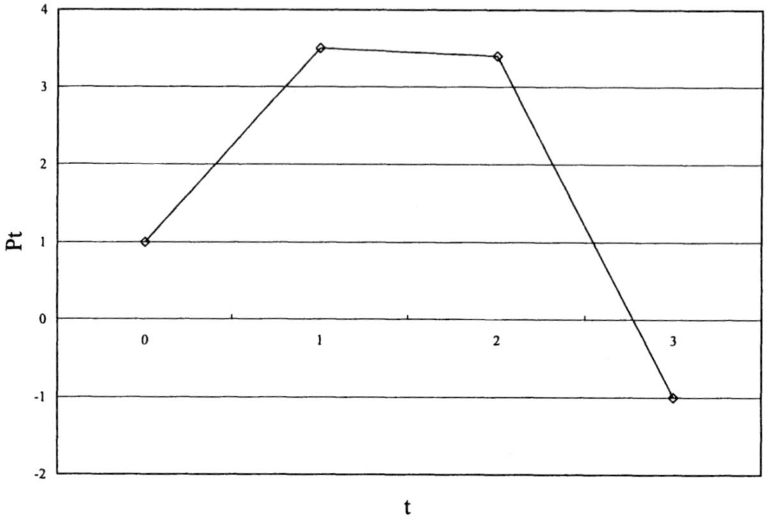

Figure 3. Price path when there is momentum trading. In this case, the price fluctuates around P_0 at period 1 and 2, and sharply falls at period 3. This observation indicates that the price process does not reflect private signals observed by rational traders. Rational traders try to earn profit by utilizing momentum trading.

We observe from the above numerical examples that the dominance of momentum investors causes a market bubble and a crash as an aftermath. De Long *et al.* [9] also obtained a similar result. However, they only focus on how and why the momentum trading causes a market bubble, but do not provide its countermeasures. Do we have a way to avert such bubble? The next numerical example presents one answer to the question.

Example 3. Simple calculations yield the relations

$$P_1 = \frac{1}{1-\kappa}P_1^{\mathrm{NM}} - \frac{\kappa}{1-\kappa}P_0$$

and

$$P_2 = P_2^{\mathrm{NM}} + \kappa(P_1 - P_0) = P_2^{\mathrm{NM}} + \frac{\kappa}{1-\kappa}(P_1^{\mathrm{NM}} - P_0), \tag{13}$$

where P_t^{NM} denotes the market price without momentum (i.e., $\phi = 0$). We observe from the above expressions that the impact of momentum trading can be decreased if $\kappa := \phi/r\rho_{v_2}$ can be reduced. For example, if we have a higher value of ρ_{v_2}, the conditions of bubbles given in Propositions 3.2 does not hold, and the price is expected to behave normally.

Note also that the coefficient of y_2 in (3) is always positive irrespective of the parameter setting. We propose that one way to solve the problem is to make the public information more precise, i.e., take a higher value of ρ_{η_2}. This means that if we know that the future public announcement such as an earnings forecast is reliable enough, then prices at both periods behave appropriately even when ϕ is relatively large.

In this numerical example, we set $\rho_{\eta_2} = 28$, and the values of the other parameters to be the same as in Example 2. Then, the parameter κ is equal to 0.25.

Figure 4 illustrates that the price path reflects private information owned by rational traders.

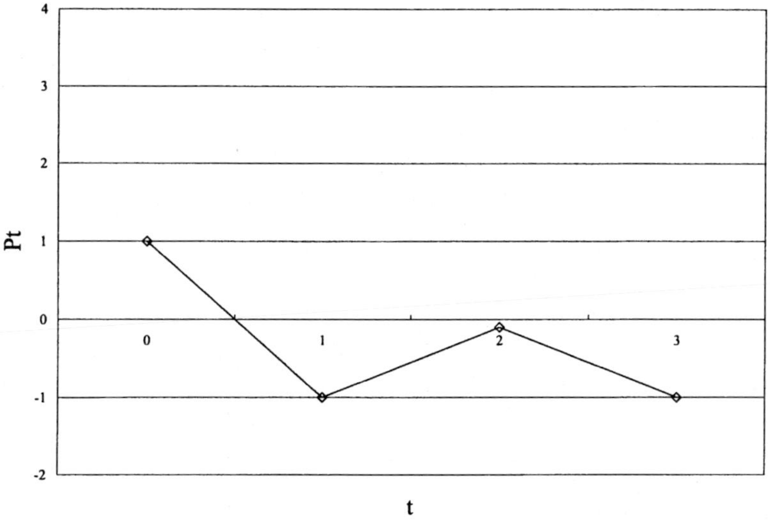

Figure 4. Price path when there is momentum trading ($\phi = 8$), but $\rho_{\eta_2} = 28$. The price reflects private information at period 1, and no bubble occurs in the market. When public announcements are reliable, rational traders trade based on their private signal rather than utilizing momentum trading. Thus, the price behaves naturally.

The movement of the price is not monotonic as in Example 1. This is because rational traders still take advantage of the momentum trading. That is, the effect of momentum trading cannot be fully erased (See Equation (13)). However, the prices at period 1 and 2 do not fluctuate sharply in comparison to Example 2, and are closer to $P_3 = v$, the liquidation value.[6] This means that when market

[6]When ρ_{η_2} is set so that ϕ is around $r\rho_{v_2}$, the demand of rational traders is insensitive to P_1 as we saw in the previous section. Therefore, halfway large ρ_{η_2} may destabilize the market price.

participants believe that there is a reliable public announcement at period 2, even the market price at period 1 properly reflects the private information owned by rational traders.

The precision of the public information affects the trading strategies of rational traders at all periods. The more reliable is the public information at period 2, the more aggressive are rational traders. The reliability of public announcements has an impact on the price via trades by rational traders. Thanks to high precision of the information on the liquidation value, rational investors trade according to the information on the liquidation value rather than utilize the existence of momentum trading. Consequently, the price reflects public and private information properly, and a market bubble does not occur.

This finding has an important implication that the market transparency is a key factor for the stable market. Financial authorities that are in charge of market design and its implementation should take into account the fact that information precision at a later period affects the price even at earlier periods.

Moreover, the results in this paper shed new light on the market momentum in the literature. Most theoretical papers, especially in behavioral finance, assume that market participants are of bounded rationality, have some bias about their own belief, or have incomplete information on the structure of the economic environment, as in Barberis *et al.* [2], Daniel *et al.* [8] and Brav and Heaton [5]. In this paper it is also assumed that there are some momentum traders in the market, but our results indicate that the price momentum depends not only on traders who are not fully rational, but also on rational investors whose strategy is consistent with the traditional expected utility theory. If it is publicly known that some of the market participants cause a price momentum, rational investors try to take advantage of their behavior, and trade the asset according to the strategy that is based on the momentum effect and are not possible if there were only rational investors in the market. Price momentum is caused not only by the momentum trades or other irrational strategies, but also by trades by rational investors who utilize the existence of anomalies. Then, public and private information observed by market participants is not properly incorporated into the price, and a large deviation of the price from the fundamentals appears in the market.

Our results also echo some of the papers that empirically study the momentum effects. For example, Chan [7] showed that the market underreacts to firm-specific news released publicly, and overreacts to news implied by price changes. Our result indicate that the finding by Chan [7] may partly be due to the existence of momentum traders and the strategy by rational traders, rather than the information update. Zhang [20] provided the evidence that information uncertainty yields price continuation anomalies. Zhang [20] attributed the incompleteness of the initial market reaction to new public news to the price momentum. We conjecture from our results that the strategy of rational investors may affect the price momentum as well as the underreaction of the news arrival, since the reliability of the public news has an impact on the order by the rational traders. The study from these point of view will be the future research.

5. Conclusions

In this paper, we have extended Kim and Kim and Verrecchia [16], studied a market where there are momentum traders who trades according to the past price movement, and analyzed how momentum trading affects the market price. Although there are some similar results to Kim and Verrecchia [16] (the first part of Proposition 3.1), the equilibrium properties significantly differ in many aspects. We have shown that the orders from momentum traders have an impact on the market price in relation to the rational traders' risk tolerance and the precision of public and private information.

When traders who trade according to the past price movement dominate the market, whether trend-chasing or contrarian, the market price becomes informationally inefficient. If there are so many trend-chasing traders in the market, the behavior of the price is abnormal, and a market bubble may occur. Our results imply that market participants or financial authorities should pay attention not only to momentum trading, but also to rational traders' trading behavior when they analyze the market. One of the ways to avoid the market instability is to improve the informational efficiency.

Acknowledgment

The author is grateful to Masaaki Kijima at Tokyo Metropolitan University, Carl Chiarella at University of Technology Sydney, Masamitsu Ohnishi at Osaka University, and seminar participants at Kyoto University for their invaluable comments and suggestions. The author also acknowledges an anonymous referee for his/her comments and suggestions. The remaining errors are, of course, of the author. The author greatly acknowledges the financial support by the Ministry of Education, Science, Sports and Culture (MEXT), "Special Coordination Funds for Promoting Science and Technology (JST-SCF)." An earlier version of this paper was circulated as Discussion paper No. 86, Graduate School of Economics, Kyoto University.

Appendix

A. Proof of Theorem 3.1

In Appendix A, we provide the detailed proof of Theorem 3.1. The procedure is similar to Appendix A in Kim and Verrecchia [16].

First, we conjecture that P_1 and P_2 are linear functions of random variables. That is, suppose that

$$P_1 = a_1 P_0 + b_1 v + c_1 x + d_1 y_1 \tag{14}$$

and

$$P_2 = a_2P_0 + b_2v + c_2x + d_{21}y_1 + d_{22}y_2 + e_2q_2, \tag{15}$$

where all coefficients are constants. Define

$$\theta_1 := v + \frac{c_1}{b_1}x = \frac{1}{b_1}(P_1 - a_1P_0 - d_1y_1) \tag{14'}$$

and

$$\theta_2 := v + \frac{c_2}{b_2}x = \frac{1}{b_2}(P_1 - a_2P_0 - d_{21}y_1 - d_{22}y_2 - e_2q_2). \tag{15'}$$

The parameter θ_t is $\mathcal{I}_{jt}$-measurable at period t because all variables in the right-hand side, including the market price P_t, are measurable when rational traders trade in the market. If $c_1/b_1 \neq c_2/b_2$, then the liquidation value v fully reveals at period 2 by solving the simultaneous equations system of (14′) and (15′). In this paper, we only consider the market price that does not fully reveal the liquidation value v at period 2. Hence, we also conjecture that $c_1/b_1 = c_2/b_2$, and that rational traders cannot know the true value of v at period 2. In the case where $c_1/b_1 = c_2/b_2$, $\theta_1 = \theta_2$ holds. We denote $\theta := \theta_1 = \theta_2$.

Now, we solve the optimization problem of rational trader j and derive the market prices backwardly.

Equilibrium at $t = 2$. Note that $\mathcal{I}_{j2}$, the information available to trader j at period 2, is the σ-field generated by s_j, y_1, y_2, P_1 and P_2. In our setting, observing P_1 and P_2 with y_1 and y_2 is statistically equivalent to observing θ. Hence, we can rewrite

$$\mathcal{I}_{j2} = \sigma(s_j, y_1, y_2, \theta).$$

Also note from (14′) and (15′) that θ is a noisy unbiased signal of v with noise term $\frac{c_2}{b_2}x$. Then, we obtain from the projection theorem of normal distributions that

$$\mathbb{E}[v|\mathcal{I}_{j2}] = \frac{\rho_{v_0}P_0 + \rho_{\epsilon_j}s_j + \rho_{\eta_1}y_1 + \rho_{\eta_2}y_2 + \rho_\theta\theta}{\rho_{v_0} + \rho_{\epsilon_j} + \rho_{\eta_1} + \rho_{\eta_2} + \rho_\theta} \tag{16}$$

and

$$\mathbb{V}[v|\mathcal{I}_{j2}] = \frac{1}{\rho_{v_0} + \rho_{\epsilon_j} + \rho_{\eta_1} + \rho_{\eta_2} + \rho_\theta} =: \frac{1}{\rho_{j2}}, \tag{17}$$

where $\rho_\theta = \left(\frac{b_2}{c_2}\right)^2 \rho_x$, the precision of $\frac{c_2}{b_2}x$.

From (16), (17), and the moment generating function of normal distributions, the optimal demand of rational trader j at period 2 is calculated as

$$\begin{aligned} x^*_{j2} :=& \frac{r_j}{\mathbb{V}[v|\mathcal{I}_{j2}]}(\mathbb{E}[v|\mathcal{I}_{j2}] - P_2) \\ =& r_j(\rho_{v_0}P_0 + \rho_{\epsilon_j}s_j + \rho_{\eta_1}y_1 + \rho_{\eta_2}y_2 + \rho_\theta\theta - \rho_{j2}P_2). \end{aligned}$$

Hence, by the demand-supply equality (1), we obtain

$$\begin{aligned} x =& \int_0^1 r_j(\rho_{v_0}P_0 + \rho_{\epsilon_j}s_j + \rho_{\eta_1}y_1 + \rho_{\eta_2}y_2 + \rho_\theta\theta - \rho_{j2}P_2)\mathrm{d}j + q_2 \\ =& r\rho_{v_0}P_0 + r\rho_\epsilon v + r\rho_{\eta_1}y_1 + r\rho_{\eta_2}y_2 + r\rho_\theta\theta - r\rho_{v_2}P_2 + q_2, \end{aligned} \tag{18}$$

where we have used $\int_j r_j\rho_{\epsilon_j}\epsilon_i\mathrm{d}j = 0$ by the law of large numbers as in Hellwig [11] or Admati [1]. Solving (18) with respect to P_2 and substituting $\theta = v + \frac{c_2}{b_2}x$ into (18) leads to

$$\begin{aligned} P_2 =& \frac{1}{\rho_{v_2}}\left(\rho_{v_0}P_0 + \rho_\epsilon v + \rho_{\eta_1}y_1 + \rho_{\eta_2}y_2 + \rho_\theta\theta - \frac{1}{r}x + \frac{1}{r}q_2\right) \\ =& \frac{\rho_{v_0}}{\rho_{v_2}}P_0 + \frac{\rho_\epsilon + \rho_\theta}{\rho_{v_2}}v + \frac{r\rho_\theta\frac{c_2}{b_2} - 1}{r\rho_{v_2}}x + \frac{\rho_{\eta_1}}{\rho_{v_2}}y_1 + \frac{\rho_{\eta_1}}{\rho_{v_2}}y_2 + \frac{1}{r\rho_{v_2}}q_2. \end{aligned} \tag{19}$$

Comparing (19) with (15), we have

$$\begin{aligned} a_2 &= \frac{\rho_{v_0}}{\rho_{v_2}}, \quad b_2 = \frac{\rho_\epsilon + \rho_\theta}{\rho_{v_2}}, \quad c_2 = \frac{r\rho_\theta\frac{c_2}{b_2} - 1}{r\rho_{v_2}}, \\ d_{21} &= \frac{\rho_{\eta_1}}{\rho_{v_2}}, \quad d_{22} = \frac{\rho_{\eta_2}}{\rho_{v_2}}, \quad e_2 = \frac{1}{r\rho_{v_2}}. \end{aligned}$$

The relation $\frac{c_2}{b_2} = \frac{r\rho_\theta\frac{c_2}{b_2} - 1}{r(\rho_\epsilon + \rho_\theta)}$ implies that $\frac{b_2}{c_2} = -r\rho_\epsilon$. Consequently, we get $\rho_0 - \left(\frac{b_2}{c_2}\right)^2\rho_x = r^2\rho_\epsilon^2\rho_x$. Substituting the coefficients and $q = \phi(P_1 - P_0)$ into (15), we finally have the price formula at period t as

$$\begin{aligned} P_2 =& \frac{\rho_{v_0}}{\rho_{v_2}}P_0 + \frac{\rho_\epsilon(1 + r^2\rho_\epsilon\rho_x)}{\rho_{v_2}}v - \frac{1 + r^2\rho_\epsilon\rho_x}{r\rho_{v_2}}x \\ &+ \frac{\rho_{\eta_1}}{\rho_{v_2}}y_1 + \frac{\rho_{\eta_2}}{\rho_{v_2}}y_2 + \frac{1}{r\rho_{v_2}}q_2 \\ =& \left(\frac{\rho_{v_0}}{\rho_{v_2}} - \kappa\right)P_0 + \frac{\rho_\epsilon(1 + r^2\rho_\epsilon\rho_x)}{\rho_{v_2}}v - \frac{1 + r^2\rho_\epsilon\rho_x}{r\rho_{v_2}}x \\ &+ \frac{\rho_{\eta_1}}{\rho_{v_2}}y_1 + \frac{\rho_{\eta_2}}{\rho_{v_2}}y_2 + \kappa P_1. \end{aligned} \tag{20}$$

Equilibrium at $t = 1$. Using the law of iterated expectations, the objective function at period 1 becomes

$$\begin{aligned}\mathbb{E}\left[\mathrm{e}^{-\frac{1}{r_j}W_j}\middle|\mathcal{I}_{j1}\right] \propto &\mathbb{E}\left[\mathrm{e}^{-\frac{1}{r_j}[(P_2-P_1)x_{j1}+(v-P_2)x^*_{j2}]}\middle|\mathcal{I}_{j1}\right]\\ =&\mathbb{E}\left[\mathbb{E}\left[\mathrm{e}^{-\frac{1}{r_j}[(P_2-P_1)x_{j1}+(v-P_2)x^*_{j2}]}\middle|\mathcal{I}_{j2}\right]\middle|\mathcal{I}_{j1}\right]\\ =&\mathbb{E}\left[\mathrm{e}^{-\frac{1}{r_j}(P_2-P_1)x_{j1}-\frac{(E[v|\mathcal{I}_{j2}]-P_2)^2}{2V[v|\mathcal{I}_{j2}]}}\middle|\mathcal{I}_{j1}\right].\end{aligned}\tag{21}$$

Rearranging (19) and noting $\theta = v - \frac{1}{r\rho_\epsilon}x$, we get

$$\begin{aligned}P_2 =& \frac{1}{\rho_{v_2}}\left[\rho_{v_0}P_0 + \rho_\epsilon\left(v - \frac{1}{r\rho_\epsilon}x\right) + \rho_\theta\theta + \rho_{\eta_1}y_1 + \rho_{\eta_2}y_2 + \frac{1}{r}q\right]\\ =& \frac{1}{\rho_{v_2}}\left(\rho_{v_0}P_0 + (\rho_\epsilon + \rho_\theta)\theta + \rho_{\eta_1}y_1 + \rho_{\eta_2}y_2 + \frac{1}{r}q_2\right).\end{aligned}\tag{22}$$

From (16) and (22), we obtain

$$\begin{aligned}\mathbb{E}[v|\mathcal{I}_{j2}] - P_2 =&\frac{\rho_{v_0}P_0 + \rho_{\epsilon_j}s_j + \rho_{\eta_1}y_1 + \rho_{\eta_2}y_2 + \rho_\theta\theta}{\rho_{j2}}\\ &- \frac{\rho_{v_0}P_0 + (\rho_\epsilon + \rho_\theta)\theta + \rho_{\eta_1}y_1 + \rho_{\eta_2}y_2 + \frac{1}{r}q_2}{\rho_{v_2}}\\ =&\frac{1}{\rho_{j2}}\left[-(\rho_{\epsilon_j} - \rho_\epsilon)P_2 + \rho_{\epsilon_j}s_j - \rho_\epsilon\theta - \frac{1}{r}q_2\right].\end{aligned}$$

Using this relation, the objective function (21) is expressed as

$$\mathbb{E}\left[\mathrm{e}^{-\frac{1}{r_j}(P_2-P_1)x_{j1}-\frac{\left[-(\rho_{\epsilon_j}-\rho_\epsilon)P_2+\rho_{\epsilon_j}s_j-\rho_\epsilon\theta-\frac{1}{r}q_2\right]^2}{2\rho_{j2}}}\middle|\mathcal{I}_{j1}\right].\tag{23}$$

The following lemma is useful for the calculation of (23).

Lemma A.1. *Let $X \sim \mathcal{N}(\mu, \sigma^2)$. Then*

$$\mathbb{E}\left[\mathrm{e}^{-aX^2+bX}\right] = \frac{\mathrm{e}^{\frac{b^2\sigma^2-2a\mu^2+2b\mu}{2(1+2a\sigma^2)}}}{1 + 2a\sigma^2}$$

for $a > -\frac{1}{2\sigma^2}$.

In (23), P_2 is the only random variable, and follows a normal conditional on $\mathcal{I}_{j1}$ under the assumption of linearity (20) as

$$\begin{aligned}\mathbb{E}[P_2|\mathcal{I}_{i1}] =& \frac{1}{\rho_{j1}\rho_{v_2}}\Big[\rho_{j2}\rho_{v_0}P_0 + \rho_{\epsilon_j}\rho_{\eta_2}s_j + \rho_{j2}\rho_{\eta_1}y_1\\ &+ (\rho_{j2}\rho_\theta + \rho_{j1}\rho_{\epsilon_j})\theta + \frac{\rho_{j1}}{r}q_2\Big]\end{aligned}\tag{24}$$

and

(25)
$$\mathbb{V}[P_2|\mathcal{I}_{i1}] = \frac{\rho_{j2}\rho_{\eta_2}}{\rho_{j1}\rho_{v_2}^2}.$$

Therefore, from Lemma A.1, we observe that (23) is proportional to

$$-\exp\left\{\frac{P_1}{r_j}x_{j1} + \frac{\left(\frac{(\rho_{\epsilon_j}-\rho_\epsilon)\left(\rho_{\epsilon_j}s_j-\rho_\epsilon\theta-\frac{1}{r}q\right)}{\rho_{j2}} - \frac{x_{j1}}{r_j}\right)\mathbb{E}_{i1}[P_2]}{2\left(1+\frac{(\rho_{\epsilon_j}-\rho_\epsilon)^2}{\rho_{j2}}\mathbb{V}_{i1}[P_2]\right)}\right.$$
$$\left. + \frac{\left\{\frac{(\rho_{\epsilon_j}-\rho_\epsilon)\left(\rho_{\epsilon_j}s_j-\rho_\epsilon\theta-\frac{1}{r}q\right)}{\rho_{j2}} - \frac{x_{j1}}{r_j}\right\}^2\mathbb{V}_{i1}[P_2]}{2\left(1+\frac{(\rho_{\epsilon_j}-\rho_\epsilon)^2}{\rho_{j2}}\mathbb{V}_{i1}[P_2]\right)}\right\}.$$

We obtain the first-order condition as

$$\begin{aligned}
x_{j1} =& r_j\left[\frac{1}{\mathbb{V}[P_2|\mathcal{I}_{i1}]}\mathbb{E}[P_1] - \frac{1+\frac{(\rho_{\epsilon_j}-\rho_\epsilon)^2}{\rho_{j2}}\mathbb{V}_{i1}[P_2]}{\mathbb{V}_{i1}[P_2]}P_1\right.\\
&\left.+\frac{(\rho_{\epsilon_j}-\rho_\epsilon)\left(\rho_{\epsilon_j}s_j-\rho_\epsilon\theta-\frac{1}{r}q\right)}{\rho_{j2}}\right]\\
=& r_j\left[\frac{\rho_{v_2}\rho_{v_0}}{\rho_{\eta_2}}P_0 + \frac{\rho_{v_2}\rho_{\eta_1}}{\rho_{\eta_2}}y_1 + \rho_{\epsilon_j}s_j + \left(\frac{\rho_{v_2}}{\rho_{\eta_2}}(\rho_\epsilon+\rho_\theta)-\rho_\epsilon\right)\theta\right.\\
&\left.+\frac{\rho_{v_1}}{r\rho_{\eta_2}}q - \left(\frac{\rho_{v_1}\rho_{v_2}}{\rho_{\eta_2}} + (\rho_{\epsilon_j}-\rho_\epsilon)\right)P_1\right]\\
=& r_j\left[\left(\frac{\rho_{v_2}\rho_{v_0}}{\rho_{\eta_2}} - \frac{\phi\rho_{v_1}}{r\rho_{\eta_2}}\right)P_0 + \frac{\rho_{v_2}}{\rho_{\eta_2}}\rho_{\eta_1}y_1\right.\\
&+\rho_{\epsilon_j}s_j + \left(\frac{\rho_{v_2}}{\rho_{\eta_2}}(\rho_\epsilon+\rho_\theta)-\rho_\epsilon\right)\theta\\
&\left.-\left(\frac{\rho_{v_1}\rho_{v_2}}{\rho_{\eta_2}} - \frac{\phi\rho_{v_1}}{r\rho_{\eta_2}} + (\rho_{\epsilon_j}-\rho_\epsilon)\right)P_1\right],
\end{aligned} \tag{26}$$

where we have substituted $q_2 = \phi(P_1 - P_0)$ in the last equation.

Now we can obtain the formula of P_1 by taking a similar procedure as that for P_2. Plugging (24) and (25) into (26), using the demand-supply equality (1), and solving the resulting equation with respect to P_1, we obtain (2). Substituting (2) into (20) yields (3). The condition $\frac{c_1}{b_1} = \frac{c_2}{b_2}$ holds in (2) and (3) and guarantees the partial revealing price. □

B. Proof of Proposition 3.1

We have another expression of P_1 as

$$(27)\qquad P_1 = \frac{\rho_{v_0}}{\rho_{v_1}}P_0 + \frac{\rho_{\eta_1}}{\rho_{v_1}}y_1 + \frac{\rho_\epsilon + r^2\rho_\epsilon^2\rho_x}{\rho_{v_1}}v - \frac{1 + r^2\rho_\epsilon\rho_x}{r\rho_{v_1}}x + \frac{1}{r\rho_{v_2}}q_2.$$

From (20) and (27), we have

$$P_2 - P_1 = \frac{\rho_{\eta_2}}{\rho_{v_2}}\left[y_2 - \frac{\rho_{v_0}P_0 + \rho_{\eta_1}y_1 + (\rho_\epsilon + r^2\rho_\epsilon^2\rho_x)v - \frac{1+r^2\rho_\epsilon\rho_x}{r}x}{\rho_{v_1}}\right].$$

On the other hand, we observe from the projection theory of normal distributions that

$$\mathbb{E}[v|\mathcal{I}_{j1}] = \frac{\rho_{v_0}P_0 + \rho_{\epsilon_j}(v + \epsilon_j) + r^2\rho_\epsilon^2\rho_x\left(v - \frac{1}{r\rho_\epsilon}x\right)}{\rho_{j1}}.$$

Now the proposition readily follows. □

C. Proofs of Propositions 3.2 to 3.4

Propositions 3.2 to 3.4 can be proved by simple but tedious algebra. □

References

1. Admati, A. R., Rational Expectations Equilibrium for Multi-Asset Securities Markets, *Econometrica*, **53**, (1985), 629–657.
2. Barberis, N., A. Shleifer, and Vishny, R., A Model of Investor Sentiment, *Journal of Financial Economics*, **49**, (1998), 307–343.
3. Barberis, N. and Thaler, R., A Survey of Behavioral Finance, *Handbook of the Economics of Finance*, (2003), 1053–1123.
4. Black, F., Noise, *Journal of Finance*, **41**, (1986), 529–541.
5. Brav, A. and Heaton, J. B., Competing Theories of Financial Anomalies, *Review of Financial Studies*, **15**, (2002), 576–606.
6. Chan, L. C., N. Jegadeesh, and Lakonishok, J., Momentum Strategies, *Journal of Finance*, **51**, (1996), 1681–1713.
7. Chan, W. S., Stock Price Reaction to News and No-news: Drift and Reversal after Headlines, *Journal of Financial Economics*, **70**, (2003), 223–260.
8. Daniel, K., Hirshleifer, D., and Subrahmanyam, A., Investor Psychology, and Security Market Under- Overreactions *Journal of Finance*, **53**, (1998), 1839–1885.
9. De Long, J. B., Shleifer, A., Summers, L. H., and Waldmann, R. J., Positive Feedback Investment Strategies and Destabilizing Rational Speculation, *Journal of Finance*, **45**, (1990), 379–395.
10. Doukas, J. A. and Petmezas, D., Acquisitions, Overconfident Managers and Self-attribution Bias, *European Financial Management*, **13**, (2007), 531–577.
11. Hellwig, M. F., On the Aggregation of Information in Competitive Markets, *Journal of Economic Theory*, **22**, (1980), 477–498.

12. Hong, H. and Stein, J. C., A Unified Theory of Underreaction, Momentum Trading, and Overreaction in Asset Markets, *Journal of Finance*, **54**, (1999), 2143–2184.
13. Hong, H., Lim, T., and Stein, J. C., Bad New Travels Slowly: Size, Analyst Coverage, and the Profitability of Momentum Strategies, *Journal of Finance*, **55**, (2000), 265–295.
14. Jegadeesh, N. and Titman, S., Returns to Buying Winners and Selling Losers: Implications for Stock Market Efficiency, *Journal of Finance*, **48**, (1993), 65–91.
15. Kausar, A. and Taffler, R. J., *Testing Behavioral Finance Models of Market Under- and Overreaction: Do They Really Work?*, (2005), working paper.
16. Kim, O. and Verrecchia, R. E., Trading Volume and Price Reaction to Public Announcements, *Journal of Accounting Research*, **29**, (1991), 302–321.
17. Kyle, A. S., Continuous Auctions and Insider Trading, *Econometrica*, **53**, (1985), 1315–1335.
18. Milgrom, P. and Stokey, N., Information, Trade and Common Knowledge, *Journal of Economic Theory*, **26**, (1982), 17–27.
19. Subrahmanyam, A., Behavioural Finance: A Review and Syntehsis, *European Financial Management*, **14**, (2007), 12–29.
20. Zhang, X. F., Information Uncertainty and Stock Returns, *Journal of Finance*, **61**, (2006), 105–137.

Investment Game with Debt Financing

Michi Nishihara* and Takashi Shibata†

*Graduate School of Economics, Osaka University
E-mail: nishihara@econ.osaka-u.ac.jp
†Graduate School of Social Sciences, Tokyo Metropolitan University
E-mail: tshibata@tmu.ac.jp

This paper investigates firm values and investment strategies (investment, coupon, and default timing) when several firms make strategic real investments with debt financing. We derive and compare the equilibrium investment strategies in three types of duopolies: (i) two symmetric firms, both of which can issue debt, (ii) two symmetric firms, only one of which (the leader) can issue debt, and (iii) a levered firm versus an unlevered firm. We show that in (iii) and in equilibrium, the levered firm always invests prior to the unlevered firm. Further, we derive the equilibrium in a oligopoly of *n* levered firms, and show that social loss increases as the number of the firms, *n*, becomes larger.

Key words: Investment game, strategic real options, debt financing, capital structure.

1. Introduction

The real options approach has become an increasingly standard framework for the investment timing decision in corporate finance (see [2]). Although the early literature on real options investigated the monopolist's investment decisions, recent studies have investigated the problem of several firms competing in the same market from a game theoretic approach (see [1] for an overview). In particular, many studies, such as [4], [7], and [21], analyze a duopoly investment game by incorporating equilibrium into a timing game with a real options approach. Other studies have concerned incomplete information between firms (e.g., [10] and [16]) and agency conflicts in a single firm (e.g., [6] and [18]).

From a different perspective, one of the most important problems in corporate finance is the derivation of the optimal capital structure. The theory of an optimal capital structure obtained through the trade-off between tax advantages and default costs, proposed by [14] in the 1950s, has subsequently been developed in more recent work, including [11] and [3]. Naturally, studies on capital structure and financing have a deep connection with those concerning the investment timing decision. However, few real options approaches focus on these matters. Noteworthy work in this area includes [12], [19], and [20]. This body of work simultaneously investigates firm value, investment timing, debt financing, and endogenous bankruptcy in a model where a firm undertakes real investment alongside the issuance of debt.

However, most of existing literature, like [12], [19], and [20], only considers a monopoly and does not reflect on the competitive situation existing between several firms. In this paper, we extend the analysis of [19] to the case where several firms attempt to preempt the market. We derive the equilibrium investment strategies in a timing game between firms that can issue debt. Through this, we clarify the effects of competition upon firm value, investment timing, debt financing, and default timing. In order to analytically derive the equilibrium, we consider the simple situation where more than one firm is not permitted to receive profit flow from the market at the same time.[1]

We reveal the effects of debt financing on strategic investment by deriving the equilibrium for the following three types of duopolies:
(i) Competition between two symmetric firms. Both firms can issue debt. This may be interpreted as the situation where each firm has its own lender.
(ii) Competition between two symmetric firms. Only the firm that makes an investment first (the leader) can issue debt, while the other firm (the follower) cannot issue debt. This may apply to the case of shortage of lenders or a credit crunch, where no bank lends money to the follower who attempts to start a project in which the leader failed before.
(iii) Competition between two asymmetric firms. Here one firm can issue debt, while the other is unlevered for exogenous reasons such as shortage of credit.

Note that in the preemptive equilibrium with competition between unlevered firms, investment takes place at the zero net present value (NPV) point (i.e., when the NPV of the investment is zero). In contrast, we show that in the equilibrium of the duopoly cases (i), (ii), and (iii), investment takes place later than zero-NPV timing and the firm value becomes positive. This results from the possibility of the leader's bankruptcy. In addition, the debt coupons the leader issues become smaller than those of the monopolist, while the firms' leverage and credit spread are unchanged.

In particular, we show that in (iii) the levered firm always wins the race. That is, the levered firm invests with debt financing prior to the unlevered firm, and obtains much larger profit than the unlevered firm. The result could explain one of the reasons why some IT venture businesses have high leverage. That is, no entrepreneur can become the successful leader without issuing debt. We also observe that the inequality (ii) < (i) < (iii) holds with respect to both the timing of investment and the value of the levered firm. The first inequality (ii) < (i) implies

[1] This assumption is essentially the same as that of [10] and [21]

that shortage of lenders (or a credit crunch) brings about an inefficient preemption race. The second inequality (i) < (iii) means that a competitor without flexibility of issuing debt is favorable.

In addition, we derive the equilibrium strategies in the competitive situation of n symmetric levered firms. As the number of the firms, n, becomes larger, investment takes places earlier and the coupons and firm value become smaller. On letting $n \to +\infty$ investment timing hastens to the zero-NPV point and the firm value decreases to 0. The similar results have been obtained in [5] who derived the equilibrium strategies in the Cournot-Nash framework instead of the preemption game in this paper. Furthermore, we investigate the social loss from the preemptive competition among firms by comparing the outcomes in the preemptive and leader–follower games. We show that the larger the number of firms, the greater the social loss.

This paper is related to recent studies [22, 23] that have exceptionally investigated both the investment and financing decision in the competitive situation. The competitive equilibrium where the output price moves between the upper and lower barriers has been investigated in [22], while the equilibrium in a duopoly has been investigated in [23]. Indeed, the model of [23] is similar to our model where the assumption that the market is small enough to be supplied by a single firm is removed. The setting by [23] is admittedly more practical than ours because it can explain a difference between leverage of the leader and the follower. However, due to its complexity, the results have not been obtained as analytically as in this paper. In addition, this paper, unlike [23], derives the equilibrium in a oligopoly and investigates the change of the equilibrium with the number of the firms.

The paper is organized as follows. Section 2 introduces the benchmark firm values and the investment strategies of the levered and unlevered monopolists. In Section 3, we derive the firm values and the investment strategies in equilibrium for the three types of duopoly (i), (ii), and (iii). In Section 4, we derive the equilibrium in oligopoly and then investigate the social loss arising from the preemptive competition between firms. Section 5 provides several numerical examples and Section 6 concludes.

2. Monopoly

2.1 Unlevered Firm

First, let us explain the setup. This paper follows the one-growth option model in [19].[2] Assume that the firm is risk-neutral and behaves in the interests of equityholders.[3] The firm with no initial assets has an option to enter a new market.

[2] [19] considers a firm with two sequentially ordered growth options in order to investigate debt overhang.

[3] Throughout this analysis, we use the terminology "equityholders" following [19]. The model does not distinguish between equityholders and entrepreneur. Hence, for the remainder of the paper, we can

The firm can choose the investment time by observing market demand $X(t)$ at time t. The firm collects a profit flow $QX(t)$ by paying a sunk cost I, where $Q(>0)$ and $I(>0)$ are constants. We assume that the firm faces a constant tax rate $\tau \in (0,1)$. For simplicity, we assume that $X(t)$ obeys the following geometric Brownian motion:

$$\text{(1)} \qquad dX(t) = \mu X(t)dt + \sigma X(t)dB(t), X(0) = x(>0),$$

where μ and $\sigma(>0)$ are constants and $B(t)$ represents the one-dimensional standard Brownian motion. The initial value $X(0) = x$ is a sufficiently small constant so that the firm has to wait for its entry condition to be met.

We now consider the unlevered firm with all-equity financing. The unlevered firm determines its investment time T by solving the following optimal stopping time problem:

$$\text{(2)} \qquad V_{ae}(x) = \sup_{T \in \mathcal{T}} \mathbb{E}[\int_T^{+\infty} e^{-rt}(1-\tau)QX(t)dt - e^{-rT}I],$$

where $\mathcal{T}$ is a set of all $\{\mathcal{F}_t\}$ stopping times ($\{\mathcal{F}_t\}$ is the usual filtration generated by $B(t)$) and r denotes the risk-free interest rate satisfying $r > \mu$. Problem (2) is reduced to

$$V_{ae}(x) = \sup_{T \in \mathcal{T}} \mathbb{E}[e^{-rT}(\Pi(X(T)) - I)],$$

where the function $\Pi(X(T))$ is defined by

$$\text{(3)} \qquad \Pi(X(T)) = \frac{1-\tau}{r-\mu}QX(T).$$

Then the optimal investment time T^i_{ae} and the firm value $V_{ae}(x)$ are easily calculated as

$$\text{(4)} \qquad T^i_{ae} = \inf\{t > 0 \mid X(t) \geq x^i_{ae}\},$$

and

$$\text{(5)} \qquad V_{ae}(x) = \left(\frac{x}{x^i_{ae}}\right)^{\beta} (\Pi(x^i_{ae}) - I).$$

(See, for example, [2] and [13]). Here, β is a positive characteristic root defined by

$$\beta = \frac{1}{2} - \frac{\mu}{\sigma^2} + \sqrt{\left(\frac{\mu}{\sigma^2} - \frac{1}{2}\right)^2 + \frac{2r}{\sigma^2}} \ (>1),$$

replace equityholders and equity value with entrepreneur and entrepreneurial value, respectively. As another way of looking at this, we may consider that in the unlevered setting the entrepreneur does not issue equity, but has the money necessary for the investment project.

and the investment trigger x^i_{ae} is

$$x^i_{ae} = \frac{\beta}{\beta - 1}\frac{I}{\Pi(1)}. \tag{6}$$

As is well-known, the investment trigger x^i_{ae} is larger than the zero-NPV trigger $x_{NPV} = I/\Pi(1)$.

2.2 Levered Firm

This subsection summarizes the results of the one-growth option case in [19]. Consider the levered firm that can issue debt with infinite maturity at investment. In the usual manner, we solve the problem backwards.

Assume that the firm has already invested at time s with market demand $X(s)$ along with issuing debt with coupon c. The equityholders (entrepreneur) have an incentive to default after the debt is in place. They choose the default time T^d so as to maximize the equity value as follows:

$$\begin{aligned} & E(X(s), c) \\ = & \sup_{\substack{T \in \mathcal{T} \\ T \geq s}} \mathbb{E}[\int_s^T \mathrm{e}^{-r(t-s)}(1-\tau)(QX(t) - c)\mathrm{d}t \mid \mathcal{F}_s]. \end{aligned} \tag{7}$$

The optimal default time T^d is $T^d = \inf\{t \geq s \mid X(t) \leq x^d(c)\}$, where the default trigger $x^d(c)$ is the function defined by

$$x^d(c) = \frac{\gamma}{\gamma - 1}\frac{r - \mu}{r}\frac{c}{Q}. \tag{8}$$

Here γ denotes a negative characteristic root defined by

$$\gamma = \frac{1}{2} - \frac{\mu}{\sigma^2} - \sqrt{\left(\frac{\mu}{\sigma^2} - \frac{1}{2}\right)^2 + \frac{2r}{\sigma^2}} \ (< 0).$$

Then, at time s the equity value $E(X(s), c)$, the debt value $D(X(s), c)$, and the firm value $V(X(s), c) = E(X(s), c) + D(X(s), c)$ are expressed as

$$E(X(s), c) = \Pi(X(s)) - \frac{(1-\tau)c}{r} - \left(\Pi(x^d(c)) - \frac{(1-\tau)c}{r}\right)\left(\frac{X(s)}{x^d(c)}\right)^\gamma \tag{9}$$

$$D(X(s), c) = \mathbb{E}[\int_s^{T^d} \mathrm{e}^{-r(t-s)}c\mathrm{d}t + \mathrm{e}^{-r(T^d - s)}(1-\alpha)\Pi(X(T^d)) \mid \mathcal{F}_s] \tag{10}$$

$$= \frac{c}{r} - \left(\frac{c}{r} - (1-\alpha)\Pi(x^d(c))\right)\left(\frac{X(s)}{x^d(c)}\right)^\gamma \tag{11}$$

$$V(X(s), c) = \Pi(X(s)) + \frac{\tau c}{r} - \left(\alpha\Pi(x^d(c)) + \frac{\tau c}{r}\right)\left(\frac{X(s)}{x^d(c)}\right)^{\gamma} \tag{12}$$

for $X(s) \geq x^d(c)$, where $\alpha(\geq 0)$ is a given constant representing the default cost. Note that the debtholders collect the entire default value, i.e., $(1-\alpha)\Pi(x^d(c))$.

The equityholders (entrepreneur) choose the investment trigger T^i and coupon c to maximize the firm value (12). That is, the problem becomes the following:

$$V_{de}(x) = \sup_{\substack{T\in\mathcal{T} \\ c(\geq 0):\mathcal{F}_T-measurable}} \mathbb{E}[e^{-rT}(V(X(T), c) - I)]. \tag{13}$$

Problem (13) can be interpreted as follows. Assume that the debtholders lend K for the debt. The equityholders' (entrepreneur's) value at investment time T is then

$$E(X(T), c) + K - I, \tag{14}$$

while the debtholders' value at T becomes

$$D(X(T), c) - K. \tag{15}$$

Since the sum of (14) and (15) is equal to $V(X(T), c) - I$, the solution of problem (13) is optimal for both the equityholders and the debtholders. The amount K determines the asset allocation between equityholders and debtholders, but in this paper we do not consider the allocation problem.[4]

Note that $\arg\max_{c\geq 0} V(X(s), c)$ becomes

$$c(X(s)) = \frac{r}{r-\mu}\frac{\gamma-1}{\gamma}\frac{QX(s)}{h} \ (>0), \tag{16}$$

for $X(s) > 0$. Here, h is a constant given by

$$h = \left[1 - \gamma\left(1 - \alpha + \frac{\alpha}{\tau}\right)\right]^{-\frac{1}{\gamma}} > 1.$$

By some calculation we can show

$$V(X(s), c(X(s))) = \psi^{-1}\Pi(X(s)), \tag{17}$$

[4]In [12] agency conflicts between equityholders and debtholders occur at the investment time because the amount K is fixed prior to investment. In contrast, such conflicts do not arise in [19] and the current analysis because the price K is adjusted through negotiation at the time of the investment. The difference between problem (13) and the "first-best" scenario in [12] is whether the coupon c is controllable.

where the function $\Pi(\cdot)$ is defined by (3) and ψ is a constant given by

$$\psi = \left[1 + \frac{\tau}{(1-\tau)h}\right]^{-1} < 1.$$

As a result, problem (13) can be rewritten as

$$V_{de}(x) = \sup_{T \in \mathcal{T}} \mathbb{E}[\mathrm{e}^{-rT}(\psi^{-1}\Pi(X(s)) - I)].$$

Thus, the optimal investment time of (13) is

$$T^i = \inf\{t > 0 \mid X(t) \geq x^i\},$$

and the optimal coupon is $c(x^i)$, where the investment trigger x^i is defined by

(18) $$x^i = \psi x^i_{ae} < x^i_{ae}.$$

Recall that x^i_{ae} is defined by (6). From (8) and (16), we have a default trigger

(19) $$x^d(c(x^i)) = x^i/h.$$

The firm value $V_{de}(x)$ at the initial time becomes

(20) $$V_{de}(x) = \left(\frac{x}{x^i}\right)^{\beta}(\psi^{-1}\Pi(x^i) - I) = \psi^{-\beta}V_{ae}(x).$$

The investment trigger x^i of the levered firm lies between the levered firm's zero-NPV trigger ψx_{NPV} and the unlevered firm's optimal trigger x^i_{ae}. Note that the unlevered firm's problem (2) corresponds to (13) with $c = 0$. Naturally, the levered firm's value (20) is greater than that of the unlevered firm (5). The leverage LV and the credit spread CS at the time of investment are calculated as

$$LV = \frac{D(x^i, c(x^i))}{V(x^i, c(x^i))}$$

(21) $$= \frac{\gamma - 1}{\gamma}\frac{\psi(1-\xi)}{h(1-\tau)}$$

and

$$CS = \frac{c(x^i)}{D(x^i, c(x^i))} - r$$

(22) $$= r\frac{\xi}{1-\xi},$$

respectively, where ξ is defined by

$$\xi = \left(1 - (1-\alpha)(1-\tau)\frac{\gamma}{\gamma - 1}\right)h^{\gamma}.$$

Note that $0 < \xi < 1$ and both (21) and (22) do not depend on the investment trigger x^i. For further details of the results concerning the levered monopolist, see the one-growth option case in [19].

3. Duopoly

This section considers the competition between two firms with complete information and focuses on the strategic investment with debt financing. Assume that each firm receives a cash flow $Q_2X(t)$ when both firms are active in the market. In order to show the essence of the firms' preemptive activities, Sections 3.1–3.3 assume $Q_2 = 0$ as in [10] and [21]. This means that the market is small enough to be supplied by a single firm. After Section 3.1 describes the benchmark case of the competition between unlevered firms, Sections 3.2 and 3.3 investigate the situation of two symmetric firms that can issue debt, and two asymmetric firms; that is, a levered firm versus an unlevered firm. Section 3.4 gives a brief comment on the general case of $Q_2 \in (0, Q)$ (negative externalities), though we are unable to present the analytical derivation.

3.1 Competition Between Unlevered Firms

This subsection provides the well-known outcome under competition between two unlevered firms (see, for example, [7]). Let $L_{ae}(X(s))$ and $F_{ae}(X(s))$ denote the expected discounted value (at time s) of a firm that enters the market first (the leader) at $X(s)$ and that of the other firm that responds optimally to the leader (the follower). It follows from $Q_2 = 0$ that the follower has no opportunity for investment. Accordingly, the follower's and leader's values become $F_{ae}(X(s)) = 0$ and $L_{ae}(X(s)) = \Pi(X(s)) - I$, respectively. In the situation where neither firm has invested, each firm attempts to invest earlier than the other in order to obtain the leader's payoff $L_{ae}(X(s))$ when the leader's payoff $L_{ae}(X(s))$ is larger than the follower's payoff $F_{ae}(X(s))$. Through preemption, each firm tries to invest at the zero-NPV point $X(s) = x_{NPV}$,[5] which is the solution of $\Pi(X(s)) - I = 0$ in equilibrium. Consequently, each firm's value becomes zero. There are no equilibriums other than the above (referred to as the preemptive equilibrium). Note that the outcome remains unchanged in the setting where n unlevered firms compete.

3.2 Competition Between Two Symmetric Firms

This section considers two types of competition between two symmetric firms with debt financing. We first consider duopoly (i), where both firms, regardless of whether they invest first, can issue debt.

As usual, we begin where one of the firms (the leader) has already invested at $X(s)$ with issuing debt with coupon c. The leader's firm value, denoted by $L_{de}(X(s), c)$, is $L_{de}(X(s), c) = V(X(s), c) - I$ because from point s the leader can obtain the monopolist's cash flow $QX(t)$ and choose the monopolist's default strategy since $Q_2 = 0$.

On the other hand, the firm value, denoted by $F_{de}(X(s), c)$, of the other firm

[5]Following many studies including [4, 21], this paper assumes that one of the firms is chosen as a leader with probability $1/2$ when the firms try to invest at the same timing. For details of the timing game, see Appendix A.

(the follower) responding optimally against the leader is calculated as follows:

$$F_{de}(X(s),c) = \left(\frac{X(s)}{x^d(c))}\right)^{\gamma} V_{de}(x^d(c))) \tag{23}$$

Eq. (23) is the value of the option to invest after the leader's bankruptcy. As in [9], we assume that a bankrupt firm leaves the market immediately. On the other hand, [23] claims that bankruptcy does not lead the exit from the market but undergoes a reorganization process (by chapter 11 in the US). It seems to depend on the firm size which assumption holds. Our assumption seems to apply to small firms such as venture businesses rather than large corporations.

Note that the follower chooses the same investment trigger (of course, *is not at the same time*), coupon, and default trigger as the monopolist, i.e., x^i, $c(x^i)$, and $x^d(c(x^i))$. Unlike $F_{ae}(X(s)) = 0$ in Section 3.1, $F_{de}(X(s),c) > 0$ holds for all $X(s), c > 0$.

Using $L_{de}(X(s),c)$ and $F_{de}(X(s),c)$, we formulate a stopping game where both firms try to preempt each other when the leader's incentive is positive, i.e., $L_{de}(X(s),c(X(s))) > F_{de}(X(s),c(X(s)))$. Define the action space of the firms as:

$$\mathcal{A} = \{(T,c) \mid T \in \mathcal{T}, c \geq 0 : \mathcal{F}_T\text{measurable}\}. \tag{24}$$

Recall that $\mathcal{T}$ denotes the set of all stopping times. If one of the firms (denoted by firm 1) chooses the strategy $(T_1,c_1) \in \mathcal{A}$ and the other (denoted by firm 2) chooses the strategy $(T_2,c_2) \in \mathcal{A}$, firms 1 and 2 expect to obtain:

$$\pi_1(T_1,c_1,T_2,c_2) = \mathbb{E}[1_{\{T_1<T_2\}}e^{-rT_1}L_{de}(X(T_1),c_1) + 1_{\{T_1>T_2\}}e^{-rT_2}F_{de}(X(T_2),c_2)$$

$$+1_{\{T_1=T_2\}}e^{-rT_1}\frac{L_{de}(X(T_1),c_1) + F_{de}(X(T_2),c_2)}{2}],$$

and

$$\pi_2(T_1,c_1,T_2,c_2) = \mathbb{E}[1_{\{T_1<T_2\}}e^{-rT_1}F_{de}(X(T_1),c_1) + 1_{\{T_1>T_2\}}e^{-rT_2}L_{de}(X(T_2),c_2)$$

$$+1_{\{T_1=T_2\}}e^{-rT_1}\frac{F_{de}(X(T_1),c_1) + L_{de}(X(T_2),c_2)}{2}],$$

respectively. Recall the assumption that one of the firms is chosen as a leader with probability $1/2$ when both firms attempt to invest at the same time; i.e., $T_1 = T_2$.

Let us derive an equilibrium of the stopping game (denoted by game G) by two firms with strategy space $\mathcal{A}$. More precisely, we find $(\tilde{T}_1,\tilde{c}_1,\tilde{T}_2,\tilde{c}_2) \in \mathcal{A}\times\mathcal{A}$ satisfying both:

$$\pi_1(\tilde{T}_1,\tilde{c}_1,\tilde{T}_2,\tilde{c}_2) = \max_{(T_1,c_1)\in\mathcal{A}} \pi_1(T_1,c_1,\tilde{T}_2,\tilde{c}_2),$$

and

$$\pi_2(\tilde{T_1}, \tilde{c_1}, \tilde{T_2}, \tilde{c_2}) = \max_{(T_2, c_2) \in \mathcal{A}} \pi_2(\tilde{T_1}, \tilde{c_1}, T_2, c_2).$$

We need the following lemma.

Lemma 3.1. Let $f(\cdot) : [a, b] \longrightarrow \mathbb{R}$ be a continuous and convex function such that $f(a) > 0$ and $f(b) < 0$. There exists a unique solution of the equation $f(x) = 0$ in the interval (a, b).

Proof. By the continuity of $f(\cdot)$, there exists at least one solution of the equation $f(x) = 0$ in the interval (a, b). Now let us show the uniqueness. Assume that there is more than one solution. We denote two of the solutions by x_1, x_2 and assume $x_1 < x_2$ without loss of generality. By $x_1 < x_2 < b$, we can take $\lambda \in (0, 1)$ satisfying $x_2 = \lambda x_1 + (1 - \lambda)b$. Then it follows from the convexity of $f(\cdot)$ that $f(x_2) \leq \lambda f(x_1) + (1 - \lambda)f(b)$. By $f(x_1) = f(x_2) = 0$, we have $f(b) \geq 0$. This contradicts the assumption of $f(b) < 0$. Thus there is no more than one solution. □

Using Lemma 3.1, we show the following proposition.

Proposition 3.1. There exists a unique x_P satisfying $L_{de}(x_P, c(x_P)) = F_{de}(x_P, c(x_P))$, i.e.,

$$\psi^{-1}\Pi(x_P) - I = h^{\gamma-\beta}\left(\frac{x_P}{x^i}\right)^{\beta}(\psi^{-1}\Pi(x^i) - I), \tag{25}$$

in the interval $(\psi x_{NPV}, x^i)$. Note that $c(\cdot)$ is defined by (16). An equilibrium of game G is $(T_L^i, c(x_P), T_L^i, c(x_P))$, where

$$T_L^i = \inf\{t > 0 \mid X(t) \geq x_P\}.$$

In equilibrium the firm value at the initial time is

$$h^{\gamma-\beta}V_{de}(x). \tag{26}$$

Proof. The function $\Pi(\cdot)$ is linear by its definition (3), and the function $(\cdot)^{\beta}$ is convex from $\beta > 1$. The function $h^{\gamma-\beta}\left(\cdot/x^i\right)^{\beta}(\psi^{-1}\Pi(x^i) - I) - (\psi^{-1}\Pi(\cdot) - I)$ is then convex and continuous. From

$$h^{\gamma-\beta}(\frac{\psi x_{NPV}}{x^i})^{\beta}(\psi^{-1}\Pi(x^i) - I) > 0$$

and

$$\psi^{-1}\Pi(\psi x_{NPV}) - I = 0$$

which holds by definition of x_{NPV}, we have

$$h^{\gamma-\beta}(\frac{\psi x_{NPV}}{x^i})^{\beta}(\psi^{-1}\Pi(x^i) - I) - (\psi^{-1}\Pi(\psi x_{NPV}) - I) > 0.$$

From

$$\begin{aligned} h^{\gamma-\beta}\left(\frac{x^i}{x^i}\right)^{\beta}(\psi^{-1}\Pi(x^i) - I) &= h^{\gamma-\beta}(\psi^{-1}\Pi(x^i) - I) \\ &< \psi^{-1}\Pi(x^i) - I \end{aligned}$$

where the last inequality follows from $h > 1$ and $\gamma - \beta < 0$, we have

$$h^{\gamma-\beta}\left(\frac{x^i}{x^i}\right)^{\beta}(\psi^{-1}\Pi(x^i) - I) - (\psi^{-1}\Pi(x^i) - I) < 0.$$

Then, by Lemma 3.1, there exists a unique x_P satisfying (25) in the interval $(\psi x_{NPV}, x^i)$.

From (17) we have:

$$L_{de}(X(s), c(X(s))) = \psi^{-1}\Pi(X(s)) - I. \tag{27}$$

It follows from (23) and (19) that:

$$\begin{aligned} F_{de}(X(s), c(X(s))) &= \left(\frac{X(s)}{x^d(c(X(s)))}\right)^{\gamma} V_{de}(x^d(c(X(s)))) \\ &= h^{\gamma} V_{de}(X(s)/h) \\ &= h^{\gamma-\beta}\left(\frac{X(s)}{x^i}\right)^{\beta}(\psi^{-1}\Pi(x^i) - I), \end{aligned} \tag{28}$$

for $X(s) \leq x^i$. By (27), (28) and the definition of x_P, we have the relationship:

$$L_{de}(X(s), c(X(s))) - F_{de}(X(s), c(X(s))) \begin{cases} < 0 & (0 < X(s) < x_P) \\ = 0 & (X(s) = x_P) \\ > 0 & (x_P < X(s) \leq x^i). \end{cases} \tag{29}$$

Let us now check:

$$\pi_1(T_L^i, c(x_P), T_L^i, c(x_P)) = \max_{(T_1,c_1)\in\mathcal{A}} \pi(T_1, c_1, T_L^i, c(x_P)). \tag{30}$$

For an arbitrary $(T_1, c_1) \in \mathcal{A}$, we can calculate as follows:

$$\begin{aligned}
&\pi_1(T_1, c_1, T_L^i, c(x_P)) \\
&= \mathbb{E}[1_{\{T_1<T_L^i\}} e^{-rT_1} L_{de}(X(T_1), c_1) + 1_{\{T_1>T_L^i\}} e^{-rT_L^i} F_{de}(x_P, c(x_P)) \\
&\quad + 1_{\{T_1=T_L^i\}} e^{-rT_1} \frac{L_{de}(x_P, c_1) + F_{de}(x_P, c(x_P))}{2}] \\
(31) \quad &\leq \mathbb{E}[1_{\{T_1<T_L^i\}} e^{-rT_1} L_{de}(X(T_1), c(X(T_1))) + 1_{\{T_1\geq T_L^i\}} e^{-rT_L^i} L_{de}(x_P, c(x_P))] \\
(32) \quad &= \mathbb{E}[1_{\{T_1<T_L^i\}} e^{-rT_1} (\psi^{-1}\Pi(X(T_1)) - I) + 1_{\{T_1\geq T_L^i\}} e^{-rT_L^i} (\psi^{-1}\Pi(x_P) - I)] \\
&\leq \sup_{T\in\mathcal{T}} \mathbb{E}[1_{\{T<T_L^i\}} e^{-rT_1} (\psi^{-1}\Pi(X(T)) - I) + 1_{\{T\geq T_L^i\}} e^{-rT_L^i} (\psi^{-1}\Pi(x_P) - I)] \\
(33) \quad &= \mathbb{E}[e^{-rT_L^i} (\psi^{-1}\Pi(x_P) - I)].
\end{aligned}$$

Here, (31) follows from (29) and the optimality of $c(\cdot)$, i.e., $L_{de}(X(s), c(X(s))) = \max_{c\geq 0} L_{de}(X(s), c)$, while (32) follows from (27). We can show (33) by:

$$\left(\frac{x}{y}\right)^\beta (\psi^{-1}\Pi(y) - I) \leq \left(\frac{x}{x_P}\right)^\beta (\psi^{-1}\Pi(x_P) - I) \quad (x \leq y \leq x_P)$$

which results from $x_P < x^i$.

On the other hand, we obtain:

$$\begin{aligned}
\pi_1(T_L^i, c(x_P), T_L^i, c(x_P)) &= \mathbb{E}[e^{-rT_L^i} \frac{L_{de}(x_P, c(x_P)) + F_{de}(x_P, c(x_P))}{2}] \\
&= \mathbb{E}[e^{-rT_L^i} L_{de}(x_P, c(x_P))] \\
(34) \qquad &= \mathbb{E}[e^{-rT_L^i} \psi^{-1}(\Pi(x_P) - I)].
\end{aligned}$$

Thus, we obtain (30) from (33) and (34). Taking account of the symmetry, we also obtain:

$$(35) \qquad \pi_2(T_L^i, c(x_P), T_L^i, c(x_P)) = \max_{(T_2,c_2)\in\mathcal{A}} \pi(T_L^i, c(x_P), T_2, c_2).$$

By (30) and (35), $(T_L^i, c(x_P), T_L^i, c(x_P))$ is an equilibrium of game G.

Finally, we can calculate as follows:

$$\begin{aligned}
\pi_i(T_L^i, c(x_P), T_L^i, c(x_P)) &= \mathbb{E}[e^{-rT_L^i} \frac{L_{de}(x_P, c(x_P)) + F_{de}(x_P, c(x_P))}{2}] \\
&= \mathbb{E}[e^{-rT_L^i} F_{de}(x_P, c(x_P))] \\
(36) \qquad &= \left(\frac{x}{x_P}\right)^\beta h^{\gamma-\beta} \left(\frac{x_P}{x^i}\right)^\beta (\psi^{-1}\Pi(x^i) - I) \\
&= h^{\gamma-\beta} \left(\frac{x}{x^i}\right)^\beta (\psi^{-1}\Pi(x^i) - I) \\
(37) \qquad &= h^{\gamma-\beta} V_{de}(x),
\end{aligned}$$

for $i = 1, 2$, where (36) and (37) follow from (28) and (20), respectively. □

The outcome in Proposition 3.1 can be interpreted as follows. In duopoly (i) each firm attempts to invest at $T_L^i = \inf\{t > 0 \mid X(t) \geq x_P\}$ and one of the firms executes the investment as a leader at time T_L^i, along with issuing debt with coupon $c(x_P)$. The leader then defaults at $T_L^d = \inf\{t > T_L^i \mid X(t) \leq x_P/h\}$. After the leader's bankruptcy, the other firm, as a follower, invests at $T_F^i = \inf\{t > T_L^d \mid X(t) \geq x^i\}$ along with issuing debt with coupon $c(x^i)$, and then defaults $T_F^d = \inf\{t > T_F^i \mid X(t) \leq x^i/h\}$.

The preemptive trigger x_P may be smaller than the unlevered firm's zero-NPV point x_{NPV}, though of course it is larger than in the levered case, ψx_{NPV}. In fact, and as presented in Section 5, for many practical parameter values we observe $x_P < x_{NPV}$. Proposition 3.1 shows that the leader has a smaller investment trigger, coupon, and default trigger than the monopolist (or follower), i.e., $x_P < x^i$, $c(x_P) < c(x^i)$ and, $x^d(c(x_P)) = x_P/h < x^d(c(x^i)) = x^i/h$. Both firms' leverage and credit spread at the investment time remain unchanged from those of the monopolist, i.e., (21) and (22), respectively. This is because even with the fear of preemption by the rival the firm can optimize its capital structure. The firm value (26) is $h^{\gamma-\beta}(< 1)$ times the levered monopolist value (20) because of the preemptive competition.

A firm's endogenous default decision generates a positive firm value, in spite of the assumption $Q_2 = 0$. This feature contrasts with earlier results in the extant literature. In [21] and [17] a leader does not always obtain a profit from the market because it takes a random development term from investment until the completion of the project. The random development term generates a positive value under competition. In [10] incomplete information about the rival firm's strategy plays a role in generating a positive value under competition.

Next, let us turn to duopoly (ii) where only the leader can issue debt. This may be interpreted as the situation where only one lender exists for the investment project. The firm value of the leader who invests at $X(s)$ with the issuance of debt with coupon c does not change from $L_{de}(X(s), c)$. On the other hand, the firm value, denoted by $F_{ae}^{de}(X(s), c)$, of the follower who takes the optimal response is reduced to the following:

$$F_{ae}^{de}(X(s), c) = \left(\frac{X(s)}{x^d(c)}\right)^{\gamma} V_{ae}(x^d(c)). \tag{38}$$

The firms attempt to preempt each other when $L_{de}(X(s), c(X(s))) > F_{ae}^{de}(X(s), c(X(s)))$.[6] We consider the stopping game $\hat{G}$ where F_{de} is replaced by

[6] This paper considers a model where the equityholders (entrepreneur) attempt to maximize the firm value as discussed in problem (13). This paper does not consider the debtholders' optimal strategy. As will be noted in Section 6, it remains an important topic for future work to analyze how the allocation between equityholders and debtholders changes with competition among entrepreneurs.

F_{ae}^{de} in the payoff functions π_i $(i = 1, 2)$ of game G. We then obtain the following proposition in duopoly (ii).

Proposition 3.2. There exists a unique solution of $L_{de}(\hat{x_P}, c(\hat{x_P})) = F_{ae}^{de}(\hat{x_P}, c(\hat{x_P}))$ in the interval $(\psi x_{NPV}, x^i)$. An equilibrium of game $\hat{G}$ is $(\hat{T_L^i}, c(\hat{x_P}), \hat{T_L^i}, c(\hat{x_P}))$, where

$$\hat{T_L^i} = \inf\{t > 0 \mid X(t) \geq \hat{x_P}\}.$$

The firm value at the initial time becomes

$$h^{\gamma-\beta} V_{ae}(x). \tag{39}$$

Proof. The proof is given in the same way as for Proposition 3.1, and therefore we show only the fact that there exists a unique solution of the equation in the interval. We calculate, for $X(s) < x^i$,

$$F_{ae}^{de}(X(s), c(X(s))) = \left(\frac{X(s)}{x^d(c(X(s)))}\right)^{\gamma} V_{ae}(x^d(c(X(s)))$$

$$= h^{\gamma-\beta}\left(\frac{X(s)}{x_{ae}^i}\right)^{\beta} (\Pi(x_{ae}^i) - I) \tag{40}$$

and

$$L_{de}(X(s), c(X(s))) = \psi^{-1}\Pi(X(s)) - I.$$

Then, by the linearity of $\Pi(\cdot)$ and $\beta > 1$, the function $F_{ae}^{de}(\cdot, c(\cdot)) - L_{de}(\cdot, c(\cdot))$ is convex and continuous in the interval. Now let us examine the signs of the values at the ends in the interval. We have

$$F_{ae}^{de}(\psi x_{NPV}, c(\psi x_{NPV})) - L_{de}(\psi x_{NPV}, c(\psi x_{NPV})) > 0$$

by

$$F_{ae}^{de}(\psi x_{NPV}, c(\psi x_{NPV})) = h^{\gamma-\beta}\left(\frac{\psi x_{NPV}}{x_{ae}^i}\right)^{\beta} (\Pi(x_{ae}^i) - I) > 0$$

and

$$L_{de}(\psi x_{NPV}, c(\psi x_{NPV})) = \psi^{-1}\Pi(\psi x_{NPV}) - I = 0.$$

We have

$$F_{ae}^{de}(x^i, c(x^i)) - L_{de}(x^i, c(x^i)) < 0$$

by

$$F_{ae}^{de}(x^i, c(x^i)) = h^{\gamma-\beta}\left(\frac{x^i}{x_{ae}^i}\right)^{\beta} (\Pi(x_{ae}^i) - I)$$

$$< \Pi(x_{ae}^i) - I \tag{41}$$

$$= \psi^{-1}\Pi(x^i) - I \tag{42}$$

$$= L_{de}(x^i, c(x^i)),$$

where (41) follows from $h > 1$, $\gamma - \beta < 0$, $0 < x^i < x^i_{ae}$, and $\beta > 1$, while (42) follows from $x^i = \psi x^i_{ae}$ and the linearity of $\Pi(\cdot)$. Thus, by Lemma 3.1 we can show the existence and the uniqueness of the solution of $L_{de}(\hat{x}_P, c(\hat{x}_P)) = F^{de}_{ae}(\hat{x}_P, c(\hat{x}_P))$ in the interval $(\psi x_{NPV}, x^i)$. □

Proposition 3.2 can be interpreted as follows. In duopoly (ii), each firm tries to invest at $\hat{T}^i_L = \inf\{t > 0 \mid X(t) \geq \hat{x}_P\}$ and one of the firms executes the investment as a leader at time $\hat{T}^i_L$ along with issuing debt with coupon $c(\hat{x}_P)$. Then the leader defaults at $\hat{T}^d_L = \inf\{t > \hat{T}^i_L \mid X(t) \leq \hat{x}_P/h\}$. After the leader's bankruptcy, the remaining firm, as a follower, invests at $\hat{T}^i_F = \inf\{t > \hat{T}^d_L \mid X(t) \geq x^i_{ae}\}$ without debt.

It can be easily checked that $F^{de}_{ae}(X(s), c(X(s))) < F_{de}(X(s), c(X(s)))$ for $X(s) > 0$. This implies that the leader has a smaller investment trigger, coupon, and default trigger than the leader in duopoly (i), i.e., $\hat{x}_P < x_P$, $c(\hat{x}_P) < c(x_P)$, and $x^d(c(\hat{x}_P)) = \hat{x}_P/h < x^d(c(x_P)) = x_P/h$. The firms' leverage and credit spread are the same as (21) and (22) in monopoly. The firm value (39) is $V_{de}(x)/V_{ae}(x)(< 1)$ times that of (26) in duopoly (i). Compared with duopoly (i), more severe preemptive competition occurs in duopoly (ii) because the leader enjoys not only a market advantage, but also the advantage of adjusting its capital structure.

The results could be interpreted as follows. When a credit crisis happens like today, few banks attempt to lend money to entrepreneurial firms even if their plans are promising. Taking account of that the follower invests after the leader's failure in the similar project, no one anticipates that there is any lender for the follower's investment. Such a situation brings about duopoly (ii), leading the preemptive competition between firms worse.

3.3 Competition Between the Levered and Unlevered Firms

This subsection considers duopoly (iii): a levered firm versus an unlevered firm that is not allowed to issue debt for some exogenous reason, such as a shortage of credit. The firm value of the levered firm that invests as a leader at $X(s)$ agrees with $L_{de}(X(s), c)$, while the firm value of the unlevered firm that responds optimally as a follower is equal to $F^{de}_{ae}(X(s), c)$ given by (38). Conversely, the firm value of the unlevered firm that invests as a leader at $X(s)$ becomes $L_{ae}(X(s))$, while the firm value of the levered firm acting as a follower is $F_{ae}(X(s)) = 0$.[7]

The levered firm has an incentive to preempt the unlevered firm for $X(s)$ satisfying $L_{de}(X(s), c(X(s))) > F_{ae}(X(s)) = 0$, i.e., $X(s) > \psi x_{NPV}$. On the other hand, the unlevered firm attempts to become the leader for $X(s)$ satisfying $L_{ae}(X(s)) > F^{de}_{ae}(X(s), c(X(s)))$. Considering this, we have the following

[7]As shown in Proposition 3.3, the levered firm always becomes a leader in equilibrium, and therefore the order is never realized.

proposition in duopoly (iii). Let $\tilde{G}$ denote the stopping game corresponding to duopoly (iii).

Proposition 3.3. There exists a unique solution $\tilde{x_P}$ of $L_{ae}(\tilde{x_P}) = F^{de}_{ae}(\tilde{x_P}, c(\tilde{x_P}))$ in the interval (x_{NPV}, x^i_{ae}). The outcomes in duopoly (iii) are classified into the following two cases.
(a) $\tilde{x_P} \leq x^i$ (preemptive equilibrium)
An equilibrium of game $\tilde{G}$ is $(\tilde{T}^i_L, c(\tilde{x_P}), \tilde{T}^i_L, c(\tilde{x_P}))$,[8] where

$$\tilde{T}^i_L = \inf\{t > 0 \mid X(t) \geq \tilde{x_P}\}.$$

The levered firm value at the initial time is equal to

$$\left(\frac{x}{\tilde{x_P}}\right)^{\beta} (\psi^{-1}\Pi(\tilde{x_P}) - I). \tag{43}$$

The unlevered firm value agrees with (39).

(b) $\tilde{x_P} > x^i$ (dominant leader-type equilibrium)
An equilibrium of game $\tilde{G}$ is $(T^i, c(x^i), \tilde{T}^i_L, c(\tilde{x_P}))$, where $(T_i, c(x_i))$ is the monopolist's strategy and $(\tilde{T}^i_L, c(\tilde{x_P}))$ is defined in (a). The levered firm value at the initial time is the same as that of the monopolist, $V_{de}(x)$, given by (20). The unlevered firm value is (39).

Proof. The proof is given in the same way as for Proposition 3.1, and therefore we show only the fact that there exists a unique solution of the equation in the interval. We have $F^{de}_{ae}(X(s), c(X(s))) = (40)$ and

$$L_{ae}(X(s)) = \Pi(X(s)) - I$$

for $X(s) < x^i$. Then, by the linearity of $\Pi(\cdot)$ and $\beta > 1$ the function $F^{de}_{ae}(\cdot, c(\cdot)) - L_{ae}(\cdot)$ is convex and continuous. We check the signs of the values at the ends in the interval. We have

$$F^{de}_{ae}(x_{NPV}, c(x_{NPV})) - L_{ae}(x_{NPV}) > 0$$

by

$$F^{de}_{ae}(x_{NPV}, c(x_{NPV})) = h^{\gamma-\beta}\left(\frac{x_{NPV}}{x^i_{ae}}\right)^{\beta} (\Pi(x^i_{ae}) - I) > 0$$

[8]Let (T_1, c_1, T_2, c_2) denote (the levered firm's strategy, the unlevered firm's strategy). We use the terminology of "equilibrium" in case (a) in the following sense. For an arbitrary $\delta > 0$, there exists an $\epsilon > 0$ such that $\tilde{\pi}_1(\tilde{T}^i_L, c(\tilde{x_P}), \tilde{T}^i_L(\epsilon), c(\tilde{x_P} + \epsilon))) > \sup_{(T_1,c_1)\in\mathcal{A}} \tilde{\pi}_1(T_1, c_1, \tilde{T}^i_L(\epsilon), c(\tilde{x_P} + \epsilon)) - \delta$ and $\tilde{\pi}_2(\tilde{T}^i_L, c(\tilde{x_P}), \tilde{T}^i_L(\epsilon), c(\tilde{x_P} + \epsilon))) = \max_{(T_2,c_2)\in\mathcal{A}} \tilde{\pi}_2(\tilde{T}^i_L, c(\tilde{x_P}), T_2, c_2)$. Here, $\tilde{\pi}_i$ is the payoff function corresponding to duopoly (iii) and $\tilde{T}^i_L(\epsilon) = \inf\{t > 0 \mid X(t) \geq \tilde{x_P} + \epsilon\}$.

and

$$L_{ae}(x_{NPV}) = \Pi(x_{NPV}) - I = 0.$$

We have

$$F_{ae}^{de}(x_{ae}^i, c(x_{ae}^i)) - L_{ae}(x_{ae}^i) < 0$$

by

$$\begin{aligned} F_{ae}^{de}(x_{ae}^i, c(x_{ae}^i)) &= h^{\gamma-\beta}\left(\frac{x_{ae}^i}{x_{ae}^i}\right)^{\beta}(\Pi(x_{ae}^i) - I) \\ &< \Pi(x_{ae}^i) - I \qquad (44) \\ &= L_{ae}(x_{ae}^i), \end{aligned}$$

where (44) follows from $h > 1$, $\gamma - \beta < 0$. Thus, by Lemma 3.1 there exists a unique solution $\tilde{x_P}$ of $L_{ae}(\tilde{x_P}) = F_{ae}^{de}(\tilde{x_P}, c(\tilde{x_P}))$ in the interval (x_{NPV}, x_{ae}^i). □

As explained in [7] and [8], three types of equilibrium may arise, namely preemptive, dominant leader-type, and joint investment-type. In (a) in Proposition 3.3 the preemptive equilibrium occurs. In this case the levered firm[9] invests at $\tilde{T_L^i} = \inf\{t > 0 \mid X(t) \geq \tilde{x_P}\}$ along with issuing debt of coupon $c(\tilde{x_P})$, and then defaults $\tilde{T_L^d} = \inf\{t > \tilde{T_L^i} \mid X(t) \leq \tilde{x_P}/h\}$. After the levered firm's bankruptcy, the unlevered firm invests at $\tilde{T_{Fa}^i} = \inf\{t > \tilde{T_L^d} \mid X(t) \geq x_{ae}^i\}$.

On the other hand, in (b) the dominant leader-type equilibrium occurs. In this case, the levered firm invests at $T^i = \inf\{t > 0 \mid X(t) \geq x^i\}$ along with issuing debt of coupon $c(x^i)$, and then defaults $T^d = \inf\{t > T^i \mid X(t) \leq x^i/h\}$. After the levered firm's bankruptcy the unlevered firm invests at $\tilde{T_{Fb}^i} = \inf\{t > T^d \mid X(t) \geq x_{ae}^i\}$.

In both cases, the levered firm that enjoys the optimal capital structure becomes the leader. The result is realistically intuitive. A venture business that pioneers a new market often has relatively high leverage at the initial stage. Although one of the major reasons is their few internal funds to cover the investment cost, Proposition 3.3 may indicate another reason. That is to say, an entrepreneur with credit enough to issue debt always wins a preemption race and appears as the leader. For quite a large τ, which leads a small x^i, condition (b) is satisfied. In (b) the levered firm is dominant owing to its substantial tax advantage over the unlevered firm.

Let us look at the investment strategies in Proposition 3.3. Note that the unlevered firm's investment trigger is the same as the unlevered monopolist's. With respect to the levered firm's strategy in (a), we can show inequalities $\hat{x_P} < \tilde{x_P} < x^i$, $c(\hat{x_P}) < c(\tilde{x_P}) < c(x^i)$, and $x^d(c(\hat{x_P})) = \hat{x_P}/h < x^d(c(\tilde{x_P})) = \tilde{x_P}/h < x^d(c(x^i)) =$

[9] By the discussion in footnote 8, we can interpret that the levered firm invests first with probability 1.

x^i/h. As in the previous propositions, the firm's leverage and credit spread at investment time are unchanged from the (21) and (22) of the monopolist. Note that the trigger $\tilde{x_P}$, unlike x_P, is always larger than the unlevered firm's zero-NPV trigger x_{NPV}. The inequality $\tilde{x_P} > x_P$ holds for most parameter values, as shown in Section 5, though it cannot be theoretically proven. In (b), the levered firm can take the best strategy, i.e., the monopolist's strategy, because of the strong tax effect.

We now consider the firm value in Proposition 3.3. In both cases, the unlevered firm must wait for the leader's bankruptcy. Due to the waiting time, the unlevered firm's value (39) becomes $h^{\gamma-\beta}(< 1)$ times the monopolist's value. Note that the unlevered firm's value is the same in both cases, in spite of $T^{\tilde{i}}_{Fa} \neq T^{\tilde{i}}_{Fb}$. The levered firm value in (a) is also reduced from that of the monopolist because of the suboptimal investment timing. By $\tilde{x_P} > x_P > \hat{x_P}$, the levered firm's value (43) becomes larger than (39) and (26) in duopolies (i) and (ii). The levered firm's value in (b) also agrees with that of the monopolist because it can take up the optimal strategy. To sum up, the fact that the rival changes from a levered to unlevered firm means a decline in the rival's competitive power and therefore an increase in the levered firm's value. Note that in both cases the levered firm's value exceeds the (39) of the unlevered firm.

3.4 Case of $Q_2 > 0$

This subsection provides a brief explanation of the results in the general case such that $0 < Q_2 < Q$, although the setting does not allow us to show clear results. Literature [23] investigates the similar case in full details with numerical examples, though bankruptcy does not mean the exit from the market in the model.

As a benchmark, we consider competition between two unlevered firms. By $Q_2 > 0$, the follower can enter the market where the leader survives when the market demand $X(s)$ is sufficiently large. The leader's profit is reduced from $QX(t)$ to $Q_2X(t)$ after the follower's entry. Since the leader's incentive is smaller than in the case of $Q_2 = 0$, the preemption trigger becomes larger than the zero-NPV trigger x_{NPV}. This generates a positive firm value in equilibrium in the case of $Q_2 \in (0, Q)$.

We now consider duopolies (i)–(iii). In every case, the follower may invest for large $X(s)$ prior to the leader's default. Note that the follower in (ii) and (iii) never defaults. This changes the leader's default trigger in the market where both are active from $x^d(c)$ to $x^d(c)Q/Q_2$. Thus, in (ii) and (iii) both the equity and debt values of the leader are reduced from (9) and (11). Expecting the possibility of the follower's interception, the leader issues debt with a smaller coupon than $c(X(s))$. On the other hand, because of the decrease in the leader's value and the increase in the follower's value, the preemption triggers, denoted by $\hat{x_P}'$ and $\tilde{x_P}'$ in (ii) and (iii) with $Q_2 \in (0, Q)$, become larger than $\hat{x_P}$ and $\tilde{x_P}$, respectively. With the trade-off between these two effects, it is ambiguous whether the leader's coupon in the investment time in (ii) and (iii) with $Q_2 \in (0, Q)$ is smaller than $c(\hat{x_P})$ and $c(\tilde{x_P})$.

The leverage and credit spread may also change from those of the monopolist.

In duopoly (i), the analysis is more complicated. The follower can choose its coupon taking account of the outcome of the exit-timing game discussed in [15] when it enters the market where the leader survives. The follower is likely to choose a smaller coupon than the leader so that it can win the exit-timing game (i.e., collect a monopolistic profit flow $QX(t)$ following the leader's bankruptcy). In this case, the leader's default trigger changes from $x^d(c)$ to $x^d(c)Q/Q_2$, which implies a similar outcome to (ii) and (iii) with $Q_2 \in (0, Q)$. The inequality $\hat{x_P}' < x_P', \tilde{x_P}'$ is unchanged, where x_P' denotes the preemption trigger in (ii) with $Q_2 \in (0, Q)$.

4. Oligopoly

4.1 Competition Among n Levered Firms

Throughout this section, we assume that the market is small enough for demand to be met by a single firm, i.e., $Q_2 = 0$. In this case, the stopping game proceeds as follows. In this section, we generalize duopoly (i) to the situation of n firms that can issue debt. In this case, the stopping game proceeds as follows:

The first game, denoted by $G_{(n)}$, is the stopping game by n firms at the initial time. In game $G_{(n)}$ one of the firms, denoted Firm n, invests as the leader and defaults. The sub-game, denoted by $G_{(n-1)}$, by the $n-1$ remaining firms starts at the default time of Firm n. In game $G_{(n-1)}$ one of the firms, denoted Firm $n-1$, invests as the leader and defaults.

⋮

The sub-game $G_{(2)} = G$ by the two remaining firms starts at the default time of Firm 3. In game $G_{(2)}$ one of the firms, denoted by Firm 2, invests as the leader and defaults.

The last firm, denoted by Firm 1, takes the monopolist's strategy after Firm 2's default.

We obtain the following proposition for games $G_{(k)}$ $(k = 1, 2, \dots, n)$.

Proposition 4.1. There exists a unique $x^i_{(k)}$ satisfying

$$\psi^{-1}\Pi(x^i_{(k)}) - I = h^{(k-1)(\gamma-\beta)} \left(\frac{x^i_{(k)}}{x^i} \right)^{\beta} (\psi^{-1}\Pi(x^i) - I) \tag{45}$$

in the interval $(\psi x_{NPV}, x^i]$ for $k = 1, 2, \dots, n$.

An equilibrium of game $G_{(k)}$ is that all the remaining firms take the same strategy $(T^i_{(k)}, c(x^i_{(k)}))$. Here, for $k = 1, 2, \dots, n$,

$$T^i_{(k)} = \inf\{t > T^d_{(k+1)} \mid X(t) \geq x^i_{(k)}\}$$

and

$$T^d_{(k)} = \inf\{t > T^i_{(k)} \mid X(t) \le x^i_{(k)}/h\},$$

where we define $T^d_{(n+1)} = 0$.

The investment triggers $x^i_{(k)}$ satisfy

$$\psi x_{NPV} < x^i_{(n)} < x^i_{(n-1)} < \ldots < x^i_{(2)} = x_P < x^i_{(1)} = x^i. \tag{46}$$

In equilibrium, the firm value at the initial time is equal to

$$h^{(n-1)(\gamma-\beta)} V_{de}(x). \tag{47}$$

As $n \to +\infty$, the firm value (47) and the preemption trigger $x^i_{(n)}$ converge to 0 and ψx_{NPV}, respectively.

Proof. We first prove that Eq. (45) has a unique solution in the interval. It is explicit that (45) has a unique solution $x^i_{(1)} = x^i$ for $k = 1$. Consider $k \ge 2$. Note that the left-hand side of (45) is linear, and the right is convex and continuous, with respect to $x^i_{(k)}$ for the interval. It follows by definition that

$$\psi^{-1}\Pi(\psi x_{NPV}) - I = 0.$$

Of course, we have

$$h^{(k-1)(\gamma-\beta)} \left(\frac{\psi x_{NPV}}{x^i}\right)^{\beta} (\psi^{-1}\Pi(x^i) - I) > 0.$$

Then, we obtain

$$\psi^{-1}\Pi(\psi x_{NPV}) - I < h^{(k-1)(\gamma-\beta)} \left(\frac{\psi x_{NPV}}{x^i}\right)^{\beta} (\psi^{-1}\Pi(x^i) - I).$$

We have

$$\begin{aligned}\psi^{-1}\Pi(x^i) - I &> h^{(k-1)(\gamma-\beta)}(\psi^{-1}\Pi(x^i) - I)\\ &= h^{(k-1)(\gamma-\beta)} \left(\frac{x_i}{x_I}\right)^{\beta} (\psi^{-1}\Pi(x^i) - I),\end{aligned}$$

where the first inequality results from $h > 1$ and $\gamma - \beta < 0$. Then, by Lemma 3.1 there exists a unique solution $x^i_{(k)} \in (\psi x_{NPV}, x^i)$. The inequality (46) follows from the feature that $h^{(k-1)(\gamma-\beta)}$ in (45) monotonically decreases with k.

Now let us solve the game backwards. The sub-game between the two firms after Firm 3's bankruptcy is the same as the game that we investigated in Proposition 3.1. By Proposition 3.1, in equilibrium both firms take the same strategy

$(T^i_L, c(x_P))$. It follows from $x_P = x^i_{(2)}$ that $(T^i_L, c(x_P)) = (T^i_{(2)}, c(x^i_{(2)}))$. Next consider the sub-game among the three firms after Firm 4's bankruptcy. The leader who invests at $X(s)$ with the issuance of debt with coupon c obtains

$$L_{de}(X(s), c) = V(X(s), c) - I, \tag{48}$$

while the two followers obtain

$$\left(\frac{X(s)}{x^d(c)}\right)^{\gamma} h^{\gamma-\beta} V_{de}(x^d(c)). \tag{49}$$

if $0 < x^d(c) < x^i_{(2)}$. Note that the followers obtain the firm value calculated in the sub-game between the two firms after the leader's (Firm 3's) default. The investment trigger $x^i_{(3)}$ defined by (45) coincides with the value where (48) with $c = c(X(s))$ is equal to (49) with $c = c(X(s))$. Then, an equilibrium is that the three firms have the same strategy $(T^i_{(3)}, c(x^i_{(3)}))$. By repeating the reasoning, we can show that An equilibrium of game $G_{(k)}$ is that all the remaining firms take the same strategy $(T^i_{(k)}, c(x^i_{(k)}))$.

We now consider the firm value (at the initial time) of Firm k in Proposition 4.1. The value can be calculated as

$$\begin{aligned}
&\left(\frac{x}{x^i_{(n)}}\right)^{\beta} h^{\gamma} \left(\frac{x^i_{(n)}}{hx^i_{(n-1)}}\right)^{\beta} h^{\gamma} \cdots \left(\frac{x^i_{(k+2)}}{hx^i_{(k+1)}}\right)^{\beta} h^{\gamma} \left(\frac{x^i_{(k+1)}}{hx^i_{(k)}}\right)^{\beta} (\psi^{-1}\Pi(x^i_{(k)}) - I) \\
&= h^{(n-k)(\gamma-\beta)} \left(\frac{x}{x^i_{(k)}}\right)^{\beta} (\psi^{-1}\Pi(x^i_{(k)}) - I) \\
&= h^{(n-k)(\gamma-\beta)} \left(\frac{x}{x^i_{(k)}}\right)^{\beta} h^{(k-1)(\gamma-\beta)} \left(\frac{x^i_{(k)}}{x^i}\right)^{\beta} (\psi^{-1}\Pi(x^i) - I) \qquad (50) \\
&= h^{(n-1)(\gamma-\beta)} V_{de}(x), \qquad (51)
\end{aligned}$$

where (50) results from the definition (15). Note that the value (51) does not depend on k — in other words, the firms are indifferent to the order of investment. The firm value in the equilibrium is equal to (51)=(47). As $n \to +\infty$, (47) $\downarrow 0$. Particularly, Firm n's value is

$$\left(\frac{x}{x^i_{(n)}}\right)^{\beta} (\psi^{-1}\Pi(x^i_{(n)}) - I) \downarrow 0 \quad (n \to +\infty),$$

which implies $x^i_{(n)} \downarrow \psi x_{NPV}$ $(n \to +\infty)$. □

Proposition 4.1 can be interpreted as follows. Each firm tries to invest at $T^i_{(n)} = \inf\{t > 0 \mid X(t) \geq x^i_{(n)}\}$ and one of the firms, denoted by Firm n, executes the investment at $T^i_{(n)}$[10] along with issuing debt with coupon $c(x^i_{(n)})$. Then Firm n defaults at $T^d_{(n)} = \inf\{t > T^i_{(n)} \mid X(t) \leq x^i_{(n)}/h\}$. After Firm n's default, the $n-1$ remaining firms attempt to invest at $T^i_{(n-1)} = \inf\{t > T^d_{(n)} \mid X(t) \geq x^i_{(n-1)}\}$ and one of the firms, denoted by Firm $n-1$ executes the investment at $T^i_{(n-1)}$ along with issuing debt with coupon $c(x^i_{(n-1)})$. Then Firm $n-1$ defaults at $T^d_{(n-1)} = \inf\{t > T^i_{(n-1)} \mid X(t) \leq x^i_{(n-1)}/h\}$. $\cdots$ After Firm 2's default, the last firm, denoted by Firm 1, invests at $T^i_{(1)} = \inf\{t > T^d_{(2)} \mid X(t) \geq x^i_{(1)}\}$ along with issuing debt with coupon $c(x^i_{(1)})$, and then defaults at $T^d_{(1)} = \inf\{t > T^i_{(1)} \mid X(t) \leq x^i_{(1)}/h\}$.

From Proposition 4.1 we have the inequalities $x^i_{(k+1)} < x^i_{(k)}$, $c(x^i_{(k+1)}) < c(x^i_{(k)})$, and $x^d(c(x^i_{(k+1)})) = x^i_{(k+1)}/h < x^d(c(x^i_{(k)})) = x^i_{(k)}/h$ (see Table 1). As in the previous propositions, the leverage and credit spread at the investment time are unchanged from those of the monopolist. The firm value (47) is $h^{(\gamma-\beta)(n-1)}(< 1)$ times the monopolist's value $V_{de}(x)$. The firm value monotonically decreases to 0 as the number of firms, n, increases. This can be interpreted as the case where a positive excess profit that arises in oligopoly (i.e., finite n) vanishes in the competitive market (i.e., infinite n). In the competitive market where an infinite number of firms compete, every firm attempts to invest at the zero-NPV trigger ψx_{NPV}. The result that the investment trigger and the firm value decrease with the number of the firms is similar to those in [5] and [10], though their models are different from ours.

4.2 Social Loss Due to Preemption

This subsection focuses on social loss from preemptive competition among firms. We first consider the outcome of the leader–follower game where the order of the firms is exogenously given in advance. Without fear of preemption by the other firms, every firm chooses the monopolist's strategy, i.e., investment trigger x^i, coupon $c(x^i)$, and default trigger $x^d(c(x^i)) = x^i/h$ (see Table 2). The firm value of Firm k, which invests after $n-k$ firms default, is derived as

$$h^{(n-k)(\gamma-\beta)}V_{de}(x). \tag{52}$$

By comparing Table 1 with Table 2, we can see the inefficiency caused by the preemption. We can see that the value of Firm 1, which is given the worst role in the leader–follower game, in Table 2 agrees with the value of all firms in the preemptive equilibrium in Table 1. The total sum of the values of n firms is

$$nh^{(\gamma-\beta)(n-1)}V_{de}(x) \downarrow 0 \quad (n \to +\infty) \tag{53}$$

in the preemptive equilibrium, while the sum in the leader–follower game is

$$\sum_{k=1}^{n} h^{(\gamma-\beta)(k-1)}V_{de}(x) = \frac{1-h^{(\gamma-\beta)n}}{1-h^{\gamma-\beta}}V_{de}(x) \uparrow \frac{V_{de}(x)}{1-h^{\gamma-\beta}} \quad (n \to +\infty). \tag{54}$$

[10] As in Proposition 3.1, we assume that one of the firms is chosen with a probability of $1/n$.

We define the (relative) social loss[11] from preemption by n firms, denoted by $Loss(n)$, as $Loss(n) = 1-(53)/(54)$. Then we have

$$(55) \qquad Loss(n) = 1 - \frac{nh^{(\gamma-\beta)(n-1)}(1-h^{\gamma-\beta})}{1-h^{(\gamma-\beta)n}} \uparrow 1 \quad (n \to +\infty).$$

From (55) we can state that an increase in the number of firms, n, causes severe preemptive competition and an inefficient outcome with greater social loss.

Table 1. Preemptive game.

	Firm n	Firm $n-1$	$\cdots$	Firm 2	Firm 1
Investment	$x^i_{(n)} <$	$x^i_{(n-1)} <$	$\cdots <$	$x^i_{(2)} = x_P <$	$x^i_{(1)} = x^i$
Coupon	$c(x^i_{(n)}) <$	$c(x^i_{(n-1)}) <$	$\cdots <$	$c(x^i_{(2)}) <$	$c(x^i_{(1)})$
Default	$x^i_{(n)}/h <$	$x^i_{(n-1)}/h <$	$\cdots <$	$x^i_{(2)}/h <$	$x^i_{(1)}/h$
Value	$h^{(n-1)(\gamma-\beta)}V_{de}(x)$	$h^{(n-1)(\gamma-\beta)}V_{de}(x)$	$\cdots$	$h^{(n-1)(\gamma-\beta)}V_{de}(x)$	$h^{(n-1)(\gamma-\beta)}V_{de}(x)$

Table 2. Leader-follower game.

	Firm n	Firm $n-1$	$\cdots$	Firm 2	Firm 1
Investment	x^i	x^i	$\cdots$	x^i	x^i
Coupon	$c(x^i)$	$c(x^i)$	$\cdots$	$c(x^i)$	$c(x^i)$
Default	x^i/h	x^i/h	$\cdots$	x^i/h	x^i/h
Value	$V_{de}(x) >$	$h^{\gamma-\beta}V_{de}(x) >$	$\cdots >$	$h^{(n-2)(\gamma-\beta)}V_{de}(x) >$	$h^{(n-1)(\gamma-\beta)}V_{de}(x)$

5. Numerical Examples

This section describes the economic implications, using some numerical examples of the theoretical results given in Sections 3 and 4. Unless otherwise noted, in what follows we set the parameters as $r = 0.07, \mu = 0.04, \sigma = 0.2, I = 2, Q = 0.1, \tau = 0.4, \alpha = 0.2, x = X(0) = 0.5$ (these parameters are typical). Then, we obtain Table 3. In Table 3, the investment triggers $x^i_{ae} = 3.291$ and $x^i = 2.4$ are 3.29 times as large as the zero-NPV triggers $x_{NPV} = 1$ and $\psi x_{NPV} = 0.729$, respectively. This means that the monopolist without rivals can wait for sufficiently large market demand $X(t)$. The levered firm chooses its optimal leverage $LV = 80.3$ % and invests earlier than the unlevered firm. The levered firm's investment threshold is $\psi = 0.729$ times as large as that of the unlevered firm. The unlevered and levered firms' values at the initial time are, respectively, $V_{ae}(x = 0.5) = 0.306$ and

[11]Strictly speaking, this is the firms' loss since this paper does not take the consumers' profits into account.

$V_{de}(x = 0.5) = 0.481$. That is, because of its optimal capital structure, the levered firm obtains 1.57 times as great a profit as the unlevered firm.

To begin with, we consider the case of a duopoly. Table 4 shows: investment triggers x^i, x_P, $\hat{x_P}$, $\tilde{x_P}$; coupons $c(x^i)$, $c(x_P)$, $c(\hat{x_P})$, $c(\tilde{x_P})$; default triggers $x^d(c(x^i))$, $x^d(c(x_P))$, $x^d(c(\hat{x_P}))$, $x^d(c(\tilde{x_P}))$; and firm values (20), (26), (39), (43). Recall that x_P, $\hat{x_P}$, and $\tilde{x_P}$ represent the preemption triggers in duopolies (i), (ii), and (iii), respectively. Note that the unlevered firm value in duopoly (iii) is equal to (39), i.e., 0.032.

We observe $\hat{x_P} < x_P < \tilde{x_P}$ and a large difference between $x_P = 0.763$ and $\tilde{x_P} = 1.046$ from the first row of Table 4. This property held for all other parameter values that we tried other than the example presented, although the inequality $x_P < \tilde{x_P}$ cannot be analytically proven. In fact, for most parameter settings, we observe that x_P and $\hat{x_P}$ are so close to the zero-NPV trigger ψx_{NPV} that $x_P < x_{NPV}$ by the assumption $Q_2 = 0$.

In particular, in duopoly (iii), we see quite a large gap between the firm values of the levered and unlevered firms, though practical values of the tax rate τ always lead to Case (a) in Proposition 3.3. In Table 4, the levered firm's value in duopoly (iii) is 0.301, which is 9.41 times as big as that of the unlevered firm, 0.032, and is 6.02 times as big as that of the levered firm in duopoly (i), 0.05. Recall that the effects of the optimized capital structure in monopoly is 1.57 times. When compared with the monopoly case, whether a firm in a duopoly can afford to issue debt is a crucial point that influences the outcome.

Table 5 presents investment triggers $x^i_{(n)}$, coupons $c(x^i_{(n)})$, default triggers $x^d(c(x^i_{(n)}))$, firm values (47), and social loss (55) under competition among n levered firms for $n = 1, 2, 3, 4$. We observe from the last two rows of Table 5, that the firm value is almost completely lost and the social loss becomes nearly 100 % for $n = 4$. This is because for most parameter values we have $h^{(\gamma-\beta)} \approx 0.1 \sim 0.2$, which means $h^{4(\gamma-\beta)} \approx 10^{-4} \sim 10^{-3}$. Then, we have $Loss(n) \approx 1, x^i_{(n)} \approx x_{NPV}$ for $n \geq 4$. The result implies that the follower's strategy of waiting up to the leaders' entries and defaults is ineffective. Indeed, the preemptive competition becomes too severe for the firms to obtain profits even if the number of the firms is quite small, i.e., $4(<< +\infty)$.

Finally, we show some interesting numerical comparative static results with respect to the volatility σ in the market demand $X(t)$. Table 6 compares $h^{\gamma-\beta}$ and $Loss(n)$ ($n = 2, 3, 4$) for $\sigma = 0.1, 0.2, 0.3, 0.4$ with other parameter values unchanged. In Table 6, $h^{\gamma-\beta}$ monotonically increases with σ, and by this $Loss(n)$ monotonically decreases with σ. This result is intuitive. For a higher volatility σ, firms actively enter and default. Then, the firm value of a follower who must wait for the leader's investment and default becomes higher, and the leader's incentive becomes lower. This moderates the firms' preemptive activities and makes the social loss smaller. The finding always held for all other parameter values that we tried other than the example presented.

Table 3. Values of quantities.

β	γ	h	ψ	x_{NPV}	ψx_{NPV}	x^l_{ae}	x^l	LV	CS
1.437	−2.437	1.796	0.729	1	0.729	3.291	2.4	80.3 %	1.32 %

Table 4. Comparison of a monopoly with duopolies (i), (ii), and (iii).

	Monopoly	Duopoly (i)	Duopoly (ii)	Duopoly (iii)
Investment	2.4	0.763	0.75	1.046
Coupon	0.44	0.14	0.137	0.192
Default	1.336	0.425	0.418	0.582
Value	0.481	0.05	0.032	0.301

Table 5. n levered firms for various n.

	Monopoly	Duopoly (i)	$n = 3$	$n = 4$
Investment	2.4	0.763	0.733	0.73
Coupon	0.44	0.14	0.134	0.134
Default	1.336	0.425	0.408	0.406
Value	0.481	0.05	0.005	0.0005
$Loss(n)$	0 %	81.3 %	97.1 %	99.6 %

Table 6. $Loss(n)$ for various σ.

	$\sigma = 0.1$	$\sigma = 0.2$	$\sigma = 0.3$	$\sigma = 0.4$
$h^{\gamma-\beta}$	0.051	0.103	0.141	0.169
$Loss(2)$	90.2 %	81.3 %	75.3 %	71.1 %
$Loss(3)$	99.3 %	97.1 %	94.9 %	92.8 %
$Loss(4)$	99.9 %	99.6 %	99 %	98.4 %

6. Conclusion

This paper investigated strategic investment with debt financing by extending the monopolist's one-growth option case in [19] to a situation allowing for the preemptive activities of several firms. We analyzed three types of duopolies, namely, (i) two symmetric firms that, whether leader or follower, can issue debt, (ii) two symmetric firms of which only the leader can issue debt, and (iii) a levered firm versus an unlevered firm. The main results for dapples can be summarized as follows.

Unlike the competition between unlevered firms, the possibility of the leader's default generates a positive excess profit to the firms in equilibrium. In (iii) the levered firm always invests first and overwhelms the unlevered firm. The order of difficulty of the preemptive competition is then (ii), (i), (iii).

Moreover, we derived the equilibrium in oligopoly of n levered firms, and showed that the social loss from preemption increases with the number of firms.

Using numerical examples, we observed that a higher volatility moderates the preemptive competition and reduces the social loss.

We now consider some interesting topics for future research. Following [19], the model in this paper does not impose any exogenous restriction on the investment cost I and the amount $K(< D(X(T), c))$ which the entrepreneur borrows by means of debt financing. In the real world, a small entrepreneurial firm that cannot issue equity is likely to have a strict restriction imposed where a part of I must be financed by debt. In a model with such a restriction we may need to reveal the effects of the competition on the leverage, although the analytical derivation of the equilibrium seems difficult. Furthermore, in a model where the debtholders are regarded as an independent player in the investment game, we may need to understand how competition among several entrepreneurs changes the allocation between entrepreneurs and debtholders.

References

1. Boyer, M., Gravel, É., and Lasserre, P., *Real options and strategic competition: a survey*, (2004), Working paper, Université de Montréal.
2. Dixit, A. and Pindyck, R., *Investment Under Uncertainty*, (1994), Princeton University Press, Princeton.
3. Goldstein, R., Ju, N., and Leland, H., An ebit-based model of dynamic capital structure, *Journal of Business*, **74**, (2001), 483–512.
4. Grenadier, S., The strategic exercise of options: development cascades and overbuilding in real estate markets, *Journal of Finance*, **51**, (1996), 1653–1679.
5. Grenadier, S., Option exercise games: an application to the equilibrium investment strategies of firms, *Review of Financial Studies*, **15**, (2002), 691–721.
6. Grenadier, S. and Wang, N., Investment timing, agency, and information, *Journal of Financial Economics*, **75**, (2005), 493–533.
7. Huisman, K., *Technology Investment: A Game Theoretic Real Options Approach*, (2001), Kluwer Academic Publishers, Boston.
8. Kong, J. and Kwok, Y., Real options in strategic investment games between two asymmetric firms, *European Journal of Operational Research*, **29**, (2007), 1405–1428.
9. Lambrecht, B., The impact of debt financing on entry and exit in a duopoly, *Review of Financial Studies*, **14**, (2001), 765–804.
10. Lambrecht, B. and Perraudin, W., Real options and preemption under incomplete information, *Journal of Economic Dynamics and Control*, **27**, (2003), 619–643.
11. Leland, H., Corporate debt value, bond covenants, and optimal capital structure, *Journal of Finance*, **49**, (1994), 1213–1252.
12. Mauer, D. and Sarkar, S., Real options, agency conflicts, and optimal capital structure, *Journal of Banking and Finance*, **29**, (2005), 1405–1428.
13. McDonald, R. and Siegel, D., The value of waiting to invest, *Quarterly Journal of Economics*, **101**, (1986), 707–727.
14. Modigliani, F. and Miller, M., The cost of capital, corporation finance and the theory of investment, *American Economic Review*, **48**, (1958), 261–297.
15. Murto, P., Exit in duopoly under uncertainty, *RAND Journal of Economics*, **35**, (2004), 111–127.

16. Nishihara, M. and Fukushima, M., Evaluation of firm's loss due to incomplete information in real investment decision, *European Journal of Operational Research*, **188**, (2008), 569–585.
17. Nishihara, M. and Ohyama, A., R&D competition in alternative technologies: A real options approach, *Journal of the Operations Research Society of Japan*, **51**, (2008), 55–80.
18. Nishihara, M. and Shibata, T., The agency problem between the owner and the manager in real investment: The bonus-audit relationship, *Operations Research Letters*, **36**, (2008), 291–296.
19. Sundaresan, S. and Wang, N., *Dynamic investment, capital structure, and debt overhang*, (2007), Working Paper, Columbia University.
20. Sundaresan, S. and Wang, N., Investment under uncertainty and strategic debt service, *American Economic Review, Papers & Proceedings*, **97**, (2007), 256–261.
21. Weeds, H., Strategic delay in a real options model of R&D competition, *Review of Economic Studies*, **69**, (2002), 729–747.
22. Zhdanov, A., Competitive equilibrium with debt, *Journal of Financial and Quantitative Analysis*, **42**, (2007), 709–734.
23. Zhdanov, A., *Optimal capital structure in a duopoly*, (2008), Working Paper Univesity of Lausanne.

The Valuation of Callable Financial Commodities with Two Stopping Boundaries

Katsushige Sawaki*, Atsuo Suzuki† and Kyoko Yagi‡

*Nanzan Business School, Nanzan University, E-mail: sawaki@nanzan-u.ac.jp
†Faculty of Urban Science, Meijo University, E-mail: atsuo@urban.meijo-u.ac.jp
‡Faculty of Economics, The University of Tokyo, E-mail: yagi@e.u-tokyo.ac.jp

In this paper we consider a model of valuing callable financial commodities which enable both an issuer and an investor to exercise their rights, respectively. We show that such a model can be formulated as a coupled stochastic game for the optimal stopping problem with two reflecting barriers. It is also shown that there exists a pair of optimal stopping rules and the value of the stochastic game. Most previous work concerning American options, Israeli options, convertible bonds and callable derivatives has required the specific payoff function when either of the issuer or the investor has exercise their option. However, we deal with a rather general payoff function of the underlying asset price and the time. We also explore some analytical properties of optimal stopping rules of the issuer and the investor.

1. Introduction

In this paper we deal with callable securities which, roughly speaking, enable both a seller (issuer) and a buyer (holder) to exercise their rights, respectively. The seller can cancel the contact by paying an extra penalty to the buyer. On the other hand, the buyer can exercise their rights at any time up to the maturity date. The game, or Israeli option, introduced by Kifer [19] is one of such securities. He formulated the price of callable securities under the setting of Dynkin's games and the hedging theory. We refer to such game-type securities as callable securities. Callable convertible bonds, interest rate swaps, callable stock options and liquid yield option notes are examples of such financial derivatives. There are many versions of callable securities including callable convertible bonds (see [33]) and callable American options [22]. The decision making related to callable securities consists of : the selection of a pair of two stopping times, the cancellation time by the seller and the exercise time by the buyer. When the seller stops at a time before the buyer does, the seller must pay to the buyer an extra penalty. On the other hand, when the buyer stops first, the buyer receives the payoff from the seller.

In this paper we formulate the valuation of such callable securities under the setting of an optimal stopping problem between the seller and the buyer. Most previous work (for examples, [7], [19], [21], [23]) has required the specific payoff function when either the seller or the buyer has exercised their options. Suzuki and Sawaki [32] considers the valuation model for perpetual game options. However, we assume in this paper that the payoff function is monotone in the underlying asset price and the time.

This paper is organized as follows. Section 2 describes a valuation model for general financial commodities that will form the basis for valuing specific financial commodities. In Section 3, we discuss several examples of specific financial commodities by introducing their concrete payoff functions. In Section 4, the analytical properties for the above general payoff function in Section 2 are explored, wherein we deal with the optimal stopping regions for the seller and the buyer, respectively. In Section 5, concluding remarks are given together with some direction for the future research.

2. Model Formulation

In this paper we apply the usual Black-Scholes setting in valuing callable financial commodities. Let the trading time horizon be given by the closed interval $[0, T]$ with the maturity T. Let $B(t)$ be the riskless asset price. Then, $B(t)$ follows the ordinary differential equation:

$$dB(t) = r(t)B(t)dt, \quad B(0) > 0, \quad r(t) \geq 0, \tag{1}$$

where $r(\cdot) > 0$ is the interest rate. Let $X(t)$ be the risky asset price. Under the risk neutral world, the dynamics of $X(t)$ follows the stochastic differential equation:

$$dX(t) = (r(t) - \delta(t))X(t)dt + \kappa(t)X(t)d\tilde{W}_t, \tag{2}$$

where $\delta(\cdot) \geq 0$ is the dividend, $\kappa(\cdot)$ is the volatility and $\tilde{W}_t$ is the standard Brownian motion under the risk neutral probability $\tilde{P}$. The stopping times for the seller (called player I) and for the buyer (called player II) are described by σ and τ, respectively. The buyer wishes to exercise the right so as to maximize the payment received from the seller. On the other hand, the seller wants to cancel the contract so as to minimize the payment to the buyer. Then, for any pair of the stopping times (σ, τ), define the payoff function as follows;

$$\begin{aligned} R_t(\sigma, \tau) = & \int_t^{\sigma \wedge \tau \wedge T} e^{\int_s^{\sigma \wedge \tau} r(u)du} c(s)ds + f(\sigma, X(\sigma)) 1_{\{\sigma < \tau\}} \\ & + g(\tau, X(\tau)) 1_{\{\tau \leq \sigma < T\}} + h(X(T)) 1_{\{\sigma \wedge \tau = T\}}. \end{aligned} \tag{3}$$

Here the functions f and g are the payoffs for the seller and the buyer, respectively. $c(\cdot)$ can be treated as the deterministic coupon payment and $h(\cdot)$ as the payment depending only on the terminal value $X(T)$. Note that the payoff function depends not only on the stopping times but on the present time t, which is the different point from Kifer [19].

Assumption 2.1.

1) *The payoff functions $f(t,x), g(t,x)$ and $h(x)$ are monotone in x for each t.*

2) *The inequations among f, g and h hold as follows: For each t*

$$f(t,x) > g(t,x) \geq h(x) \quad \forall\, x \geq 0.$$

If $\sigma = \tau$, the buyer has priority over the seller. When $\sigma \wedge \tau = \min(\sigma, \tau) = T$, the payoff is assumed to be $h(X(T))$.

Remark 2.1.

1) If it is optimal for the seller not to cancel before the maturity, callable securities may be reduced to the usual non-callable one which is an American type.

2) If it is optimal for both the seller and the buyer not to exercise before the maturity, securities are reduced to the European type.

3) Even if it is optimal for the buyer not to exercise before maturity, the seller still faces the problem of selecting an optimal stopping time σ.

Theorem 2.1. *For $X(t) = x$, define*

$$\overline{V}(t,x) = \operatorname*{ess\,inf}_{\sigma \in \mathcal{T}_{t,T}} \operatorname*{ess\,sup}_{\tau \in \mathcal{T}_{t,T}} \tilde{E}\left[e^{-\int_t^{\sigma\wedge\tau} r(s)ds} R_t(\sigma,\tau) \Big| X(t) = x \right] \tag{4}$$

and

$$\underline{V}(t,x) = \operatorname*{ess\,sup}_{\tau \in \mathcal{T}_{t,T}} \operatorname*{ess\,inf}_{\sigma \in \mathcal{T}_{t,T}} \tilde{E}\left[e^{-\int_t^{\sigma\wedge\tau} r(s)ds} R_t(\sigma,\tau) \Big| X(t) = x \right]. \tag{5}$$

Then, this game possesses the value which is given by

$$V(t,x) = \overline{V}(t,x) = \underline{V}(t,x), \quad 0 \leq t \leq T. \tag{6}$$

Moreover, the optimal stopping times for the seller and the buyer are respectively given by

$$\hat{\sigma}_t = \inf\left\{ \sigma \geq t \;:\; V(\sigma, X(\sigma)) = f(\sigma, X(\sigma)) + \int_t^{\sigma} e^{\int_s^{\sigma} r(u)du} c(s)ds \right\} \wedge T,$$

$$\hat{\tau}_t = \inf\left\{ \tau \geq t \;:\; V(\tau, X(\tau)) = g(\tau, X(\tau)) + \int_t^{\tau} e^{\int_s^{\tau} r(u)du} c(s)ds \right\} \wedge T. \tag{7}$$

The proof can be carried out in the similar way to Kifer [19] and so be omitted.

Corollary 2.1. *The value function $V(t,x)$ satisfies*

$$g(t,x)+\int_0^t e^{\int_s^t r(u)du}c(s)ds \le V(t,x) \le f(t,x)+\int_0^t e^{\int_s^t r(u)du}c(s)ds,\ \forall(t,x)\in[0,T]\times\mathbf{R}^+.$$

The proof immediately follows from (4) and (5).

Theorem 2.2. (Perpetual Financial Commodities) *In addition to Assumption 2.1, if the inequality*

$$\lim_{t\to\infty}\left(f(t,x)+\int_0^t e^{\int_s^t r(u)du}c(s)ds\right)\le M_x,\quad \forall x\ge 0 \tag{8}$$

holds, then there exists the limit $V_\infty(x)\equiv\lim_{t\to\infty}V(t,x)$ which satisfies equation (6) in Theorem 2.1.

Corollary 2.2. *If $\delta(t)=0$ and $g(t,x)$ is convex and increasing in x for each t, then*

$$V(t,x)=\operatorname*{ess\,inf}_{\sigma\in\mathcal{T}_{t,T}}\tilde{E}\left[e^{-\int_t^\sigma r(u)du}R_t(\sigma,T)\mid X(t)=x\right],$$

that is, the coupled optimal stopping problem above in Theorem 2.1 can be reduced to the one only for the issuer (player I).

Proof. Since $e^{-\int_0^t r(s)ds}X(t)$ is the martingale with respect to $\tilde{P}$, the increasing and convex function $g(t,x)$ of x is a submartingale. It can be shown from the Optimal Sampling Theorem that

$$\begin{aligned}
V(t,x) &= \operatorname*{ess\,inf}_{\tau}\operatorname*{ess\,sup}_{\sigma}\tilde{E}\left[e^{-\int_t^{\sigma\wedge\tau}r(s)ds}R_t(\sigma,\tau)\Big|X(t)=x\right]\\
&= \operatorname*{ess\,inf}_{\sigma}\tilde{E}\Bigg[e^{-\int_t^\sigma r(u)du}\Bigg(\int_t^\sigma e^{\int_s^\sigma r(u)du}c(s)ds\\
&\quad +f(\sigma,X(\sigma))1_{\{\sigma<T\}}+h(X(T))1_{\{\sigma=T\}}\Bigg)\Bigg|\,X(t)=x\Bigg]\\
&= \operatorname*{ess\,inf}_{\sigma}\tilde{E}\left[e^{-\int_t^\sigma r(u)du}R_t(\sigma,T)\,\Big|\,X(t)=x\right].
\end{aligned}$$

□

3. Some Examples

In this section we discuss several examples of the callable financial commodities with the two stopping boundaries for players I and II. We suppose $r(t)=r$, $\delta(t)=\delta$ and $\kappa(t)=\kappa$.

A. European call option

Let K be the exercise price.

$$c(t)=0,\ f(t,x)=\infty,\ g(t,x)=0,\ h(x)=\max(x-K,0)=(x-K)^+,$$
$$V(t,x)=xe^{-\delta(T-t)}\Phi(d_1^+(T-t,x,K))-Ke^{-r(T-t)}\Phi(d_1^-(T-t,x,K)),$$

where $\Phi(\cdot)$ is the standard normal cumulative distribution function and

(9) $$d_1^{\pm}(t,x,y)=\frac{\log(x/y)+(r-\delta\pm\frac{1}{2}\kappa^2)t}{\kappa\sqrt{t}}.$$

Lower bound $V(t,x)\geq 0$

B. American put option

$$c(t)=0,\ f(t,x)=\infty,\ g(t,x)=h(x)=(K-x)^+,$$
$$V(t,x)=\sup_{\tau}\tilde{E}\left[e^{-r(\tau-t)}R_0(\infty,\tau)\mid X(t)=x\right]$$

Lower and upper bounds $(K-x)^+\leq V(t,x)\leq K$

C. Game put option (Kifer [19], Suzuki and Sawaki [32])

Assume $\delta=0$ and let $p\geq 0$ be the penalty paid by the player I.

$$c(t)=0,\ f(t,x)=(K-x)^++p,\ g(t,x)=h(x)=(K-x)^+,$$
$$V(t,x)=\sup_{\tau}\inf_{\sigma}\tilde{E}\left[e^{-r(\sigma\wedge\tau-t)}R_0(\sigma,\tau)\mid X(t)=x\right]$$
$$=\inf_{\sigma}\sup_{\tau}\tilde{E}\left[e^{-r(\sigma\wedge\tau-t)}R_0(\sigma,\tau)\mid X(t)=x\right]$$

Lower and upper bounds $(K-x)^+\leq V(t,x)\leq(K-x)^++p$

The optimal stopping region for the issuer:

Let $V^{ap}(t,x)$ be the price of the American put and define

$$t^*\equiv\sup\{t\geq 0\mid p\leq V^{ap}(t,K)\}.$$

$$\mathcal{S}_t=\begin{cases}\{K\}, & t\in[0,t^*],\\ \phi, & t\in(t^*,T].\end{cases}$$

If $p>V^{ap}(t,K)$, $t^*=0$, that is, $\mathcal{S}_t=\phi$.

The optimal region for the investor

$$\mathcal{S}_t=[0,x_t]$$

Let x_t^{ap} be the optimal boundary of the American put.

$$x_t^{ap}\leq x_t\leq K \text{ and } \lim_{t\to T}x_t=K.$$

We summarize the main results (their proofs should be refered to [29]) as follows;

Theorem 3.1. *Let* $V^{ep}(t,x)$ *be the price of the European put.* $V(t,x)$ *can be decomposed as follows;*

$$V(t,x) = V^{ep}(t,x) + e(t,x) - d(t,x),$$
$$e(t,x) = rK\int_t^T e^{-r(s-t)}\Phi(d_2(s-t,x,x_s))ds \geq 0,$$
$$d(t,x) = \tilde{E}\left[\int_t^{t^*} e^{-r(s-t)}\left(\frac{\partial V}{\partial x}(s,K+) - \frac{\partial V}{\partial x}(s,K-)\right)dL_s^x(K)\Big|\, X(t)=x\right] \geq 0,$$

where $L_t^x(K)$ *is the local time of* X_t *at the level* K *in the time interval* $[0,t]$.

Corollary 3.1. *Let* $e^{ap}(t,x)$ *be the early exercise premium of the American put. Then, we obtain*

$$e(t,x) > e^{ap}(t,x), \quad t \in [0,t^*],$$
$$e(t,x) = e^{ap}(t,x), \quad t \in (t^*,T].$$

Corollary 3.2. *Let* $V_\infty^{ap}(x)$ *be the price of the perpetual American put and* x_{ap}^* *its optimal boundary and define* $p^* \equiv V_\infty^{ap}(K)$.

For $p \geq p^*$,

$$V_\infty(x) = V_\infty^{ap}(x) = rK\int_0^\infty e^{-rt}\Phi(d_2(t,x,x_{ap}^*))dt.$$

For $p < p^*$,

$$V_\infty(x) = V_\infty^{ap}(x) - \left(\frac{dV_\infty}{dx}(K+) - \frac{dV_\infty}{dx}(K-)\right)\tilde{E}\left[\int_0^\infty e^{-rt}dL_t^x(K)\Big|\, X(0)=x\right].$$

D. Callable convertible bond (Yagi and Sawaki [33] [34])

Let F be the face value, C the call price and z the dilution factor.

$$c(t) = 0,\ f(t,x) = \max(zx, C),\ g(t,x) = zx,\ h(x) = \min(x, \max(zx, F))$$

Lower and upper bounds : $zx \leq V(t,x) \leq \max(zx, C)$,

Optimal stopping boundaries

issuer

$$\mathcal{S}^I = \{(t,x)|V(t,x) = \max(zx,C)\}$$
$$x_t^I = \inf\{x|x \in \mathcal{S}_t^I\}$$
$$\mathcal{S}_t^I = [x_t^I, \infty)$$

investor

$$\mathcal{S}^{II} = \{(t,x)|V(t,x) = zx\}$$
$$x_t^{II} = \inf\{x|x \in \mathcal{S}_t^{II}\}$$
$$\mathcal{S}_t^{II} = [x_t^{II}, \infty)$$

Letting $x_t^* \equiv \min(x_t^I, x_t^{II})$, the continuing region is given by

$$C_t = [0, x_t^*).$$

Theorem 3.2. *$V(t,x)$ can be written as*

$$V(t,x) = B(t,x) + V^{ec}(t,x) + p(t,x) - d(t,x),$$

where $B(t,x)$ is the discount bond value, $V^{ec}(t,x)$ the price of European call, $p(t,x)$ the early conversion premium and $d(t,x)$ the callable discount.

The proof should be refered to [33]. Sîrbu et al [30], [31] have also considered the valuation model of callable convertible bonds under the setting of Kifer's game option. In particular, they derive the analytical solution for the case of perpetual ones, which is different from the case of the finite lived ones [33] [34].

E. Installment American call option (Ben [2], Kimura [20])
Let q be the continuous installment rate.

$$c(t) = -q,\ f(t,x) = \infty,\ g(t,x) = h(x) = (x-K)^+,$$
$$R_t(\tau_e,\tau_s) = (X(\tau_e)-K)^+ 1_{\{\tau_e<\tau_s<T\}} + (X(T)-K)^+ 1_{\{\tau_e\wedge\tau_s\geq T\}} - \int_t^{\tau_e\wedge\tau_s} e^{r(\tau_e\wedge\tau_s - s)} q\,ds,$$
$$V(t,x;q) = \operatorname*{ess\,sup}_{\tau_e,\tau_s} \tilde{E}\left[e^{-r(\tau_e\wedge\tau_s - t)} R_t(\tau_e,\tau_s) \mid X(t) = x\right]$$

Optimal stopping boundary

$$\mathcal{S} = \{(t,x)|V(t,x;q) = 0\}$$
$$\mathcal{E} = \{(t,x)|V(t,x;q) = (x-K)^+\}$$

Especially, at the maturity the optimal stopping and exercise boundaries, $\underline{x}_t$, $\overline{x}_t$ are as follows,

$$\underline{x}_T = K,$$
$$\overline{x}_T = \max\left(\frac{rK-q}{\delta}, K\right).$$

Theorem 3.3.

$$
\begin{aligned}
V(t,x;q) = V^{ec}(t,x) - q\int_t^T e^{-r(u-t)}\Phi(d_1^-(u-t,x,\underline{x}_u))du \\
+ \int_t^T \Big\{\delta x e^{-\delta(u-t)}\Phi(d_1^+(u-t,x,\overline{x}_u)) \\
-(rK-q)e^{-r(u-t)}\Phi(d_1^-(u-t,x,\overline{x}_u))\Big\}du.
\end{aligned}
$$

The proof can be refered in [20].

F. European double barrier equity linked bond

Let U be the upper barrier and L lower barrier.

$$
\begin{aligned}
c(t) &= 0,\ f(t,x) = \infty,\ g(t,x) = 0, \\
\hat{X}(T) &= \max_{0\le t\le T} X(t),\ \check{X}(T) = \min_{0\le t\le T} X(t), \\
h(\hat{X}(T),\check{X}(T)) &= F1_{\{\hat{X}(T)\ge U\}} + F1_{\{\hat{X}(T)<U,\check{X}(T)\ge L\}} \\
&\quad + \min\left(F, \frac{X(T)}{X(0)}F\right)1_{\{\hat{X}(T)<U,\check{X}(T)<L\}}, \\
V(t,x) &= \tilde{E}\Bigg[e^{-r(T-t)}\Bigg(F - \frac{F}{X(0)}(X(0)-X(T))^+1_{\{\hat{X}(T)<U\}} \\
&\quad + \frac{F}{X(0)}(X(0)-X(T))^+1_{\{\hat{X}(T)<U,\check{X}(T)\ge L\}}\Bigg)\ \Bigg|\ X(t) = x\Bigg] \\
&= e^{-r(T-t)}F - \frac{F}{X(0)}\hat{V}(t,x) + \frac{F}{X(0)}\check{V}(t,x),\ L\le x\le U,
\end{aligned}
$$

where $\hat{V}$ is the price of the European up-and-out put option with the strike price $X(0)$ and $\check{V}$ one of the European double barrier knock-out put option with the strike price $X(0)$.

G. PRDC

Let $X(t)$ be the exchange rate which follows the stochastic differential equation

$$
dX(t) = (r - r_f)X(t)dt + \kappa X(t)d\tilde{W}_t,
$$

where r is the domestic riskless interest rate and r_f the riskless rate of the counterpart.

$$
c(t) = c,\ f(t,x) = \infty,\ g(t,x) = 0,\ h(\hat{X}(T)) = \left(\alpha\frac{X(T)}{X(0)} - \beta\right)^+ F1_{\{\hat{X}(T)\le U\}},
$$

where $\alpha, \beta \ge 0$.

4. Analytical Properties

In this section we explore some analytical properties that are well known for the American put. It is of interest to investigate whether such properties still hold for the general monotone payoff function given by (3). In particular, the optimal stopping region and timing for the issuer to call the financial commodity are analyzed in more detail. To derive these properties we need the following assumption.

Assumption 4.1. *$f(t,x)$, $g(t,x)$ and $h(x)$ are non-increasing in t and monotone convex in x and $c(t) = 0$. Define the stopping regions for player I and II, respectively, by*

$$\mathcal{S}^I = \{(t,x)|V(t,x) = f(t,x)\}$$
$$\mathcal{S}^{II} = \{(t,x)|V(t,x) = g(t,x)\}$$

and the continuing region is

$$C = \{(t,x)|g(t,x) < V(t,x) < f(t,x)\} \tag{10}$$

$\mathcal{S}_t^I$ and $\mathcal{S}_t^{II}$ are the truncations of $\mathcal{S}^I$ and $\mathcal{S}^{II}$ at time t.

Lemma 4.1.

1) *$V(t,x)$ is non-increasing in t for each x.*

2) *$V(t,x)$ is monotone convex in x for each t.*

3) *If $(t,x) \in C$, then $\mathcal{L}V = 0$ where $\mathcal{L} = \frac{1}{2}\kappa^2 x^2 \frac{\partial^2}{\partial x^2} + (r-\delta)x\frac{\partial}{\partial x} + \frac{\partial}{\partial t} - r$.*

The proof can be carried out similar to the case of callable game options (see [21] [22] [29]).

Lemma 4.2. *If $f(t,x)$ and $g(t,x)$ are monotone and convex in x, then $\mathcal{S}_t^I$ and $\mathcal{S}_t^{II}$ are connected sets, that is, the stopping region never possesses the detached region.*

Proof. Assume that the stopping region $\mathcal{S}^I$ is not a connected set. For t fixed, we choose $x_1, x_2, x_1 \neq x_2, x_1, x_2 \in \mathcal{S}_t^I$. Since $V(t,x)$ is convex in x for each t, and $g(t,x)$ and $f(t,x)$ are both monotone and convex in x from Asuumption 4.1 , the set C given by (10) is convex in x for each t fixed. For α satisfying $0 < \alpha < 1$, there exists $y \in C$ such that $y = \alpha x_1 + (1-\alpha)x_2 \in C$. Then, we obtain from the definition of the continuing region

$$V(t,y) < f(t,y). \tag{11}$$

Since $V(t,x)$ is convex in x, we have from Lemma 4.1

$$\begin{aligned} V(t,y) = V(t,\alpha x_1 + (1-\alpha)x_2) &\geq \alpha V(t,x_1) + (1-\alpha)V(t,x_2) \\ &= \alpha f(t,x_1) + (1-\alpha) f(t,x_2) \\ &\geq f(t,\alpha x_1 + (1-\alpha)x_2) \\ &= f(t,y) \end{aligned}$$

which contradicts (11). □

Lemma 4.3. *If the issuer does not hold the call right, it is reduced to be American type. Let $V^a(t,x)$ be its price and, $V^a_\infty(x)$ the value of the perpetual financial commodity. We have*

1) $V(t,x) \leq V^a(t,x) \leq V^a_\infty(x)$,

2) $\mathcal{S}^{II}_t \supseteq \mathcal{S}^a_t \supseteq \mathcal{S}^a_\infty$.

Proof.

1) Since the issuer has no right to cancel the contact, the stopping problem turns out to be the stopping problem only for the investor as a maximizer. So, we have $V(t,x) \leq V^a(t,x)$ from Corollary 2.1.

2) From property 1) we obtain $\mathcal{S}^{II}_t \supseteq \mathcal{S}^a_t \supseteq \mathcal{S}^a_\infty$. □

Theorem 4.1. *Let $p_t \equiv f(t,x) - g(t,x)$, the difference depending only on t and assume that p_t is non-increasing in t and put $K_t \equiv \arg\min_x f(t,x)$. Define*

$$t^* = \sup\{t \geq 0 | p_t \leq V^a(t,K_t)\}.$$

If $t \in [0,t^*]$, the optimal call region for the issuer is $\mathcal{S}^I_t = \{K_t\}$. If $t \in (t^*,T]$, $\mathcal{S}^I_t = \phi$.

Proof. Suppose that $p_t > V^a(t,K_t)$. The issuer dose not wish to call the contract due to providing the arbitrage opportunity to the investor. Then, it is optimal for the issuer not to call the contract. Suppose $p_t \leq V^a(t,K_t)$ at time t. The minimum of the payoff function for the issuer is attained at K_t. When $t \in [0,t^*]$, we have $\mathcal{S}^I_t = \{K_t\}$. If $t \in (t^*,T]$, $V(t,x) = V^a(t,x)$. □

5. Conclusion

In this paper we have studied the valuation of callable financial commodities under the setting of the coupled optimal stopping problem. The fusion of the multiplicity of demands requested by issuers and investors creates various hybrid securities such as derivatives, callable convertible bonds and many structured products, price of which depends upon the underlying asset price. The issuer is also protected because the issuing company holds the rights to call the security back as long as the investor does not exercise his option. This contractual feature of the callable security provides the upper bound of the company's issuing cost.

We have shown that the value of the callable financial commodity possesses similar analytical properties to callable game options. In particular, the optimal stopping boundaries for both of the issuer and investor are analyzed as well as the optimal call time for the issuer. Although our model has been formulated as a two person stopping game, it is possible to extend it to a three person stopping game wherein the third player could be the stake holder such as existing stock holders. In this modeling, we must take account of the possibility of coalitions between players consisting of the issuer, investor and stake holder. We shall discuss the asset valuation model from this point of view as future work.

Acknowledgment

This paper has been presented at Daiwa International workshop on Financial Engineering in Kyoto, August 2008. The authors are grateful to Professor H. Iwaki, Kyoto University and would like to thank the referee for his/her valuable comments which make the paper more suitable for reading. This paper was supported by the Grant in Aid for Science Research(B) (20330068) and (A)(20241037) from the Japanese Society of the Promotion of Science.

References

1. Barraquand, J. and Pudet, T., Pricing of American Path-dependent Contingent Claims, *Mathematical Finance*, **6**, (1996), 17–51.
2. Ben-Ameur, H., M. Breton and P. François, Pricing Installment Options with an Application to ASX Installment Warrants, (2002), Working Paper G-2002-59, GERAD.
3. Black, F. and Scholes, M., The Pricing of Options and Corporate Liabilities, *Journal of Political Economy*, **81**, (1973), 637–659.
4. Brennan, M. J. and Schwartz, E. S., The Valuation of American Put Options, *Journal of Finance*, **32**, (1976), 449–462.
5. Carr, P., Jarrow, R. and Myneni, R., Alternative Characterizations of American Put Options, *Mathematical Finance*, **2**, (1992), 87–106.
6. Cox, J. C., Ross, S. A. and Rubinstein, M., Option Pricing: A Simplified Approach, *Journal of Financial Economics*, **7**, (1979), 229–263.
7. Cvitanic, J. and Karatzas, I., Backward Stochastic Differential Equations with Reflection and Dynkin Games, *Annals of Probability*, **24**, (1996), 2024–2056.
8. Duffie, D. and Epstein, L., Stochastic Differential Utility, *Econometrica*, **60**, (1992), 353–394.
9. Dynkin, E. B., Game Variant of a Problem on Optimal Stopping, *Soviet Mathematics Doklady*, **10**, (1969), 270–274.
10. Geman, H. and Yor, M., The Valuation of Double-Barrier Options: A Probabilistic Approach, *Mathematical Finance*, **6**, (1996), 365–378.
11. Geske, R. and Johnson, H. E., The American Put Option Valued Analytically, *Journal of Finance*, **39**, (1984), 1511–1524.

12. Ingersoll, J. E., A Contingent-Claims Valuation of Convertible Securities, *Journal of Financial Economics*, **4**, (1977), 289–322.
13. Jacka, S., Optimal Stopping and the American Put, *Mathematical Finance*, **1**, (1991), 1–14.
14. Jaillet, D., Lamberton, D. and Lapeyre, B., Variational Inequalities and the Pricing of American Options, *Acta Applicandae Mathematicae*, **21**, (1990), 263–289.
15. Karatzas, I., On the Pricing of American Options, *Applied Mathematics and Optimization*, **60**, (1988), 37–60.
16. Karatzas, I., *Lectures on the Mathematics of Finance*, (1997), CRM Monograph Series, American Mathematical Society.
17. Karatzas, I. and Shreve, S. E., *Brownian Motion and Stochastic Calculus*, (1991), Springer, New York.
18. Karatzas, I. and Shreve, S. E., *Methods of Mathematical Finance*, (1998), Springer.
19. Kifer, Y., Game Options, *Finance and Stochastics*, **4**, (2000), 443–463.
20. Kimura, T. and Kikuchi, K., Valuing continuous Time-installment Options: A Laplace Transform Approach, *Proceedings of the Sapporo Symposium on Financial Engineering and its Applications*, September (2006), Hokkaido University.
21. Kühn, C. and Kyprianou, A. E., Israeli Options as Composite Exotic Options, (2004), Preprint.
22. Kunita, H. and Seko, S., Game Call Options and Their Exercise Regions, *Technical Report of the Nanzan Academic Society Mathematical Sciences and Information Engineering*, (2004), NANZAN-TR-2004-6.
23. Kyprianou, A. E., Some Calculations for Israeli Options, *Finance and Stochastics*, **8**, (2004), 73–86.
24. McConnell, J. and Schwartz, E. S., LYON Taming, *Journal of Finance*, **41**, (1986), 561–576.
25. Merton, R. C., Theory of Rational Option Pricing, *Bell Journal of Economics and Management Science*, **4**, (1973), 141–183.
26. Myneni, R., The Pricing of the American Option, *The Annals of Applied Probability*, **2**, (1992), 1–23.
27. Pardoux, E. and Peng, S., Adapted Solution of a Backward Stochastic Differential Equation, *Systems Control Lett.*, **14**, (1990), 55–61.
28. Sawaki, K., Optimal Exercise Policies for Call Options and Their Valuation, *Computers and Mathematics with Applications*, **24**, (1992).
29. Seko, S. and Sawaki, K., The Valuation of Callable American Options, *Gendai Finance*, **17**, (2005), 83–96 (in Japanese).
30. Sîrbu, M. I. Pikovsky and S. E. Shreve, Perpetual Convertible Bonds, *SIAM J. Control Optim.*, **43**, (2004), 58–85.
31. Sîrbu, M. I. Pikovsky and S. E. Shreve, A Two-person Game for Pricing Convertible Bonds, *SIAM J. Control Optim.*, **45**, (2006), 1508–1539.
32. Suzuki, A. and Sawaki, K., The Pricing of Perpetual Game Put Options and Optimal Boundaries, *Recent Advances in Stochastic Operations Research*, (2007), World Scientific, Singapore, 175–188.
33. Yagi, K. and Sawaki, K., The Valuation and Optimal Strategies of Callable Convertible Bonds, *Pacific Journal of Optimization*, Yokohama Publishers, **1**, (2005), 375–386.
34. Yagi, K. and Sawaki, K., On the Valuation and Optimal Boundaries of Convertible Bonds with Call Notice Periods, *Recent Advances in Stochastic Operations Research*, (2007), World Scientific, Singapore, 189–202.

Statistical Properties of Covariance Estimator of Microstructure Noise: Dependence, Rare Jumps and Endogeneity

Masato Ubukata and Kosuke Oya

Graduate School of Economics, Osaka University
E-mail: ubukata@econ.osaka-u.ac.jp, oya@econ.osaka-u.ac.jp

This paper studies impacts of rare jumps as well as an endogenous microstructure noise on Ubukata and Oya's [16] cross and auto covariance estimators of bivariate microstructure noise processes. The theoretical results show that biases of the noise covariance estimators are asymptotically negligible even when there exist the jump component and the endogeneity in the observed price process. Monte Carlo results are suggestive of robustness of the noise covariance estimators to the jumps and the endogenous noise with an empirically reasonable magnitude of the correlation in finite sample.
Key words: Microstructure noise, Dependence, Rare jump, Endogeneity, High-frequency data

1. Introduction

The wide availability of high-frequency data stimulates an interest in statistical research on estimation of microstructure noise as well as integrated (co)variance of semimartingale price processes in frictionless markets. Microstructure noise can be induced by various market frictions such as the discreteness of price changes, bid-ask bounces, and asymmetric information across traders, inter alia. The recent literature on microstructure noise finds out to what extent it generates deviation of transaction prices from equilibrium prices.

The variance estimation of the noise have been proposed by Zhang, Mykland and Aït-Sahalia [18], Hansen and Lunde [10] and Bandi and Russell [4]. Bandi and Russell [3] propose an estimator of full-information transaction costs which represent the difference between transaction prices and unobserved theoretical asset prices reflecting all private and public information. The estimator is an useful tool in market microstructure literature to analysis an impact of pre-trade transparency on market quality (see Eom, Ok and Park [8]). Hansen and Lunde's [10] empirical analysis reveals some facts that the noise is quite small, time dependent and possibly correlated with the equilibrium returns and the properties of the noise change over time using thirty Dow Jones Industrial Average stocks traded on the NYSE and NASDAQ markets. Medahhi [14] examines the correlation between noise and the integrated variance. For the dependence in microstructure noise, Ubukata and Oya [16] propose a method of estimation and testing of the dependence.

It is also widely known that financial markets sometimes generate significant discontinuities called "Jumps" in financial variables and jump-diffusion process is frequently used to model high-frequency financial data. The studies of jumps detection provide the empirical evidences on the importance of jumps (see Huang and Tauchen [12], Barndorff-Nielsen and Shephard [6], Fan and Wang [9] and Lee and Mykland [13]). Meanwhile, the statistical properties of some prevailing estimators for integrated variance and noise estimation are derived under no jumps. As an example, the integrated variance of semimartingale price process is traditionally estimated using sums of squared returns over one day. This estimator, called daily realized volatility, under jump-diffusion process consists of the integrated variance and the quadratic variation of the jump component (see the discussion in Andersen, Bollerslev, Diebold and Labys [1]). In contrast, Barndorff-Nielsen and Shephard [5] propose realized bipower variation which is an integrated variance estimator robust to the presence of jumps.

Thus, it is important to see how the statistical results of the prevailing estimator change when the observed price process has the rare jumps. This paper demonstrates the impacts of the rare jumps as well as the endogenous noise on Ubukata and Oya's [16] cross and auto covariance estimators of bivariate noise processes which are derived under a diffusion process without jumps and an independence assumption of the equilibrium price and the noise. We show that the bias of the noise covariance estimator is asymptotically negligible even when there exist the jump component and the endogeneity in the observed price process because the bias term of the noise covariance estimators induced by the presence of the jumps and endogenous noise disappears as sampling frequency increases. Further, we evaluate the finite sample performance of the noise covariance estimators using Monte Carlo simulation. Monte Carlo results are suggestive of robustness of the noise covariance estimator to the jumps. We find the influence of endogenous noise with an empirically reasonable magnitude of the correlation is also negligible in finite sample.

The remainder of the paper is organized as follows. In Section 2, we will set out some notations and recall the definitions of noise cross and auto covariance estimators and the related test statistics proposed by Ubukata and Oya [16]. We study how the noise covariance estimators behave when we introduce the jumps and the endogenous noise in Section 3. In Section 4, we examine the finite sample performance of the noise covariance estimators through Monte Carlo simulation. Section 5 concludes.

2. Definitions and Previous Works

This section defines some notations for modeling high-frequency transaction price process and microstructure noise, and introduces the noise covariance estimator and the related test statistic proposed by Ubukata and Oya [16].

2.1 Notation and Microstructure Noise

Let a two-dimensional vector of logarithmic equilibrium prices for assets A and B be $\boldsymbol{P}^*(t) = (P_A^*(t), P_B^*(t))'$. It is assumed that $\boldsymbol{P}^*(t)$ follows a stochastic volatility process:

$$\boldsymbol{P}^*(t) = \int_0^t \boldsymbol{\Sigma}(s)d\boldsymbol{W}(s), \tag{1}$$

where the instantaneous or spot covolatility process $\boldsymbol{\Sigma}$ has elements that are bounded and $\boldsymbol{W}$ is a vector standard Brownian motion.

Transaction prices are deviated from equilibrium prices by the presence of microstructure noise which may be due to bid-ask spreads, the discreteness of prices change and the difference in trade sizes, and so on. The presence of microstructure noise poses a problem that the transaction prices may not be a semimartingale as showed in many empirical results for integrated variance estimation (See Hansen and Lunde [10], Mykland and Zhang [15] and Zhang, Mykland and Aït-Sahalia [18] for the details). It is natural to consider that for discrete time points t_{j_A} and t_{j_B} of assets A and B, the observed high-frequency transaction price is a sum of the logarithmic equilibrium price and the microstructure noise, that is:

$$P_y(t_{j_y}) = P_y^*(t_{j_y}) + u_y(t_{j_y}), \quad y = A, B, \tag{2}$$

where $P_y(t_{j_y})$, $P_y^*(t_{j_y})$ and $u_y(t_{j_y})$ are observed transaction prices, equilibrium prices and microstructure noises at time t_{j_y}.

We denote the sequence of transaction times of assets A and B by $\{t_{j_A} : j_A = 1, 2, \ldots, n_A\}$ and $\{t_{j_B} : j_B = 1, 2, \ldots, n_B\}$, where n_A and n_B are the number of trades for assets A and B. We assume the first trades of both assets start at $t_0 = 0$. The observed returns of assets A and B are defined as $r_{j_A} = P_A(t_{j_A}) - P_A(t_{j_A-1})$ and $r_{j_B} = P_B(t_{j_B}) - P_B(t_{j_B-1})$. It is noted that the transaction times of asset A and B are not necessarily equidistant and synchronous.

Let a vector of microstructure noise be $\boldsymbol{\Gamma}(\ell) = (u_A(t), u_B(t))'$. We assume that the bivariate noise processes are covariance stationary, with the autocovariance function having a finite dependence in the sense that:

$$\boldsymbol{\Gamma}(\ell) = \mathrm{E}[\boldsymbol{u}(t)\boldsymbol{u}'(t-\ell)] = \boldsymbol{0}, \quad \text{for all } \mid \ell \mid > m, \tag{3}$$

where m is an unknown positive number which is determined through the test statistic introduced in the following. $\boldsymbol{\Gamma}(\ell)$ is a covariance matrix with finite elements. The noise process is assumed to be independent of the equilibrium price process. In Section 3.2 the independence assumption is dropped.

2.2 Noise Cross and Autocovariance Estimators

In this section we introduce Ubukata and Oya's [16] cross and auto covariance estimators of bivariate noise processes. The elements of $\mathbf{\Gamma}(\ell)$ are expressed as:

$$\mathbf{\Gamma}(\ell) = \mathrm{E}[\boldsymbol{u}(t)\boldsymbol{u}'(t-\ell)] = \begin{pmatrix} \gamma_A(\ell) & \gamma_{AB}(\ell) \\ \gamma_{BA}(\ell) & \gamma_B(\ell) \end{pmatrix} = \begin{pmatrix} \gamma_A(\ell) & \gamma(\ell) \\ \gamma(-\ell) & \gamma_B(\ell) \end{pmatrix},$$

where $\gamma_A(\ell)$ and $\gamma_B(\ell)$ are autocovariances of noise for asset A and B, and $\gamma(\ell)$ and $\gamma(-\ell)$ are cross-covariances of noises between different assets.

2.2.1 Noise Cross-Covariance Estimator

For estimation of noise cross-covariance $\gamma(\ell)$ with $\ell > 0$, the product of nonoverlapping returns with the distance ℓ for different assets is defined as:

$$(4) \qquad Z_{\ell,j_A j_B} = r_{j_A} r_{j_B}, \quad \text{for all } j_A, j_B, \text{ such that } t_{j_A-1} - t_{j_B} = \ell.$$

In what follows, an expectation E[·] and variance V[·] are used as the ones conditional on the stochastic arrival times. From the independence increments of equilibrium price process and the exogenous noise assumption, the conditional expectation of $Z_{\ell,j_A j_B}$ is expressed as:

$$(5) \qquad \begin{aligned} \mathrm{E}[Z_{\ell,j_A j_B}] &= \mathrm{E}[u_A(t_{j_A})u_B(t_{j_B})] - \mathrm{E}[u_A(t_{j_A})u_B(t_{j_B-1})] \\ &\quad - \mathrm{E}[u_A(t_{j_A-1})u_B(t_{j_B})] + \mathrm{E}[u_A(t_{j_A-1})u_B(t_{j_B-1})] \\ &= \gamma(\ell + \Delta t_{j_A}) - \gamma(\ell + \Delta t_{j_A} + \Delta t_{j_B}) - \gamma(\ell) + \gamma(\ell + \Delta t_{j_B}), \end{aligned}$$

where $\Delta t_{j_A} := t_{j_A} - t_{j_A-1}$ and $\Delta t_{j_B} := t_{j_B} - t_{j_B-1}$. It is clear in estimation of $\gamma(\ell)$ that we only have to modify the product of returns $Z_{\ell,j_A j_B}$ such that three cross-covariances except $\gamma(\ell)$ become zero. If we can obtain the prior knowledge of a threshold value of noise cross-sectional dependence where noise cross-covariance becomes zero, this modification becomes possible. Thus, Ubukata and Oya [16] start with deriving a test statistic to determine the threshold value statistically.

Noise Cross-Covariance Estimator: Ubukata and Oya [16]

In construction of the test statistic for the threshold value selection, the sample mean of $Z_{\ell,j_A j_B}$ in (4) is used. The observations $Z_{\ell,j_A j_B}$ are arranged in ascending order of index j_A as $\{Z_{\ell,k}\}_{k=1}^{N_\ell}$. And then the sample mean of $Z_{\ell,k}$; that is $\bar{Z}_{\ell,N_\ell} := \frac{1}{N_\ell}\sum_{k=1}^{N_\ell} Z_{\ell,k}$, is used to estimate $\mathrm{E}[Z_{\ell,k}]$. Let $\omega_\ell^2 := \lim_{N_\ell\to\infty}\left(\bar{Z}_{\ell,N_\ell} - \mathrm{E}[\bar{Z}_{\ell,N_\ell}]\right)^2 N_\ell$ be an asymptotic variance of standardized $\bar{Z}_{\ell,N_\ell}$. For the variance estimation, the subsampling method proposed by Carlstein [7] is applied, that is, $\{Z_{\ell,k}\}_{k=1}^{N_\ell}$ is divided into nonoverlapping subseries that consist of M_ℓ consecutive observations.

The h-th subseries can be defined as follows:

$$\{Z_\ell^h\} := (Z_{\ell,hM_\ell+1}, Z_{\ell,hM_\ell+2}, \cdots, Z_{\ell,(h+1)M_\ell}),\quad 0 \le h \le K_\ell - 1,\quad K_\ell = [N_\ell/M_\ell],$$

where $[\,\cdot\,]$ denotes the integer part of a real number, M_ℓ is a number of observations within each subseries $\{Z_\ell^h\}$ and K_ℓ denotes the total number of subseries. The variance estimator can be expressed as:

$$\hat{\omega}_\ell^2 = \frac{M_\ell}{K_\ell} \sum_{h=0}^{K_\ell-1} \Big(\bar{Z}_{\ell,M_\ell}^h - \frac{1}{K_\ell} \sum_{h=0}^{K_\ell-1} \bar{Z}_{\ell,M_\ell}^h\Big)^2,$$

where $\bar{Z}_{\ell,M_\ell}^h$ is the sample mean of subseries $\{Z_\ell^h\}$. The test statistic for the threshold value selection proposed in Ubukata and Oya [16] is as follows.

Theorem 2.1. (Test Statistic: Ubukata and Oya [16])
Let the null hypothesis be $E[Z_{\ell,k}] = 0$ for all k, for given ℓ while the alternative hypothesis consists of all possible deviations from the null. As N_ℓ goes to infinity s.t. $M_\ell \to \infty$ and $M_\ell/N_\ell \to 0$, then we have:

$$\tau(\ell) := \frac{\sqrt{N_\ell}\bar{Z}_{\ell,N_\ell}}{\hat{\omega}_\ell} \xrightarrow{a} N(0,1) \tag{6}$$

under the null hypothesis. $\tau(\ell)$ diverges under the alternative.

Proof. See Ubukata and Oya [16]. □

Procedure for the Threshold Value Selection
We can identify the threshold value of noise cross-sectional dependence using the following testing procedure. First, we start with testing the null hypothesis $E[Z_{L,k}] = 0$ for all k with a large value L. The null would not be rejected because the cross-covariance between sufficiently separated noise terms is zero. If $E[Z_{L,k}] = 0$ is not rejected as expected, then we test whether $E[Z_{L-1,k}]$ is zero. This test is reiterated sequentially until the null is rejected. Finally, the distance for which the null is rejected for the first time can be determined as the threshold value of cross-sectional dependence.

Once we determine the threshold value, the noise cross-covariance estimator can be derived. Some additional notation is needed in order to introduce the noise cross-covariance estimator. Suppose that the bivariate noise processes have finite cross-sectional dependence in the sense that $\gamma(\ell) = 0$ for $\ell > m^+ > 0$ and for $-\ell > m^- > 0$. When ℓ is positive, it is possible to define $\bar{t}_{j_A}^{(+)}$ as the first transaction time for asset A, which follows t_{j_A} subject to $\bar{t}_{j_A}^{(+)} - t_{j_B} > m^+$. Likewise, $\underline{t}_{j_B-1}^{(+)}$ can be defined as the last transaction time for asset B, which is followed by t_{j_B-1} subject to

$t_{j_A-1} - \underline{t}^{(+)}_{j_B-1} > m^+$. As ℓ is negative, $\overline{t}^{(-)}_{j_B}$ refers to the first transaction time for asset B, which follows t_{j_B} subject to $\overline{t}^{(-)}_{j_B} - t_{j_A} > m^-$ while $\underline{t}^{(-)}_{j_A-1}$ is the last transaction time for asset A, which is followed by t_{j_A-1} subject to $t_{j_B-1} - \underline{t}^{(-)}_{j_A-1} > m^-$. The returns on the extended intervals $(t_{j_A-1}, \overline{t}^{(+)}_{j_A}]$ and $(\underline{t}^{(-)}_{j_A-1}, t_{j_A-1}]$ are denoted by $\overline{r}^{(+)}_{j_A}$ and $\underline{r}^{(-)}_{j_A}$, respectively. Similarly for asset B, the returns on the intervals $(t_{j_B-1}, \overline{t}^{(-)}_{j_B}]$ and $(\underline{t}^{(+)}_{j_B-1}, t_{j_B}]$ are denoted by $\overline{r}^{(-)}_{j_B}$ and $\underline{r}^{(+)}_{j_B}$, respectively. Then, we construct $Z^{(\pm)}_{\ell,j_A j_B}$ as follows:

$$Z^{(\pm)}_{\ell,j_A j_B} = \begin{cases} \overline{r}^{(+)}_{j_A} \underline{r}^{(+)}_{j_B} 1_{\{t_{j_A-1}-t_{j_B}=\ell\}} & \text{if } \ell > 0 \\ \overline{r}^{(+)}_{j_A} \underline{r}^{(+)}_{j_B} 1_{\{t_{j_A-1}-t_{j_B}=0\}} + \underline{r}^{(-)}_{j_A} \overline{r}^{(-)}_{j_B} 1_{\{t_{j_B-1}-t_{j_A}=0\}} & \text{if } \ell = 0 \\ \underline{r}^{(-)}_{j_A} \overline{r}^{(-)}_{j_B} 1_{\{t_{j_B-1}-t_{j_A}=-\ell\}} & \text{if } \ell < 0 \end{cases}$$

where $1_{[\cdot]}$ is an indicator function. We select $Z^{(\pm)}_{\ell,j_A j_B}$ such that the indicator function is equal to one. The ranking of the selected $Z^{(\pm)}_{\ell,j_A j_B}$ in ascending order of index j_A results in the sequence $\{Z^{(\pm)}_{\ell,k}\}_{k=1}^{N^{(\pm)}_\ell}$. The consistent cross-covariance estimator of $\gamma(\ell)$ given in Ubukata and Oya [16] is:

$$\hat{\gamma}(\ell) = -\frac{1}{N^{(\pm)}_\ell} \sum_{k=1}^{N^{(\pm)}_\ell} Z^{(\pm)}_{\ell,k}. \tag{7}$$

The test statistic for statistical significance of noise cross-covariances $\gamma(\ell)$ is also derived in Corollary 1 of Ubukata and Oya [16].

2.2.2 Noise Autocovariance Estimator

Construction of autocovariance estimator of univariate noise process is similar to the case of cross-covariance estimator, that is, we have only to use the product of nonoverlapping returns by univariate transaction price series.

In the same way noise cross-covariance estimator is done, we set out constructing a test statistic to determine a threshold value of univariate noise dependence using the following nonoverlapping returns:

$$Z_{\ell,i_A j_A} = r_{i_A} r_{j_A}, \qquad \text{for all } i_A, j_A, \text{ such that } t_{j_A-1} - t_{i_A} = \ell \geq 0.$$

Let $\{Z_{A,\ell,k}\}_{k=1}^{N_{A,\ell}}$ be the sequence of $Z_{\ell,i_A j_A}$, which is arranged in ascending order of the index i_A. The total number of observations in the sequence $\{Z_{A,\ell,k}\}$ is denoted as $N_{A,\ell}$. For a given ℓ, the null hypothesis is $\mathrm{E}[Z_{A,\ell,k}] = 0$ for all k, which implies that $\gamma_A(\ell) = 0$. The alternative hypothesis consists of all possible deviations from the null. Ubukata and Oya [16] derives the test statistic for the null as follows:

$$\tau_\eta(\ell) := \frac{\sqrt{N_{A,\ell}}\, \bar{Z}_{\ell,N_{A,\ell}}}{\hat{\omega}_{A,\ell}}, \tag{8}$$

where $\bar{Z}_{\ell,N_{A,\ell}}$ is the sample mean of $\{Z_{A,\ell,k}\}$. $\hat{\omega}^2_{A,\ell}$ is a subsampling estimator of $\omega^2_{A,\ell} = \lim_{N_{A,\ell}\to\infty} \mathrm{E}[(\bar{Z}_{\ell,N_{A,\ell}} - \mathrm{E}[\bar{Z}_{\ell,N_{A,\ell}}])^2 N_{A,\ell}]$. Under the null, $\tau_A(\ell) \xrightarrow{a} N(0,1)$ as $N_{A,\ell}$ goes to infinity s.t. $M_{A,\ell} \to \infty$ and $M_{A,\ell}/N_{A,\ell} \to 0$.

Using the threshold value (denoted as m_A) determined by the test statistic (8), the consistent noise autocovariance estimator is given as follows:

$$\hat{\gamma}_A(\ell) = -\frac{1}{N^{(\pm)}_{A,\ell}} \sum_{k=1}^{N^{(\pm)}_{A,\ell}} Z^{(\pm)}_{A,\ell,k}, \tag{9}$$

where $\{Z^{(\pm)}_{A,\ell,k}\}_{k=1}^{N^{(\pm)}_{A,\ell}}$ is a sequence of $Z^{(\pm)}_{\ell,i_A j_A}$ arranged in ascending order of the index i_A. $Z^{(\pm)}_{\ell,i_A j_A}$ is denoted as:

$$Z^{(\pm)}_{\ell,i_A j_A} = \underline{r}^{(-)}_{i_A} \bar{r}^{(+)}_{j_A} = \big(P_A(t_{i_A}) - P_A(\underline{t}^{(-)}_{i_A-1})\big)\big(P_A(\bar{t}^{(+)}_{j_A}) - P_A(t_{j_A-1})\big), \tag{10}$$
$$\text{for all } i_A, j_A, \text{ such that } t_{j_A-1} - t_{i_A} = \ell \geq 0,$$

where $\bar{t}^{(+)}_{j_A}$ is the first transaction time, which immediately follows t_{j_A} subject to $\bar{t}^{(+)}_{j_A} - t_{i_A} > m_A$, and $\underline{t}^{(-)}_{i_A-1}$ is the last transaction time, which precedes t_{i_A-1} subject to $t_{j_A-1} - \underline{t}^{(-)}_{i_A-1} > m_A$. See Ubukata and Oya [16] for the details of the noise autocovariance estimator and the test statistics for its statistical significance.

The cross and auto covariance estimators introduced in this section are derived under no jumps and exogenous noise assumptions. In the following sections, we will see how these estimators and test statistics are affected under the jump diffusion process and the endogenous noise.

3. Rare Jumps and Endogenous Noise

We study the statistical properties of the estimator when the equilibrium price process is contaminated with the rare jumps and endogenous noise.

3.1 Rare Jumps and Noise Covariance Estimator

Define the transaction price $P(t_{j_y})$ as:

$$P_y(t_{j_y}) = P^*(t_{j_y}) + J_y(t_{j_y}) + u_y(t_{j_y}), \quad y = A, B \tag{11}$$

where $P^*(t_{j_y})$ and $u_y(t_{j_y})$ are the equilibrium prices and the noises at time t_{j_y}. The jump variable J_y is assumed to follow a compound Poisson process, which is most commonly employed in the literature, given by:

$$J_y(t_{j_y}) = \sum_{i=1}^{N^J_y(t_{j_y})} c_{y,i}, \quad y = A, B \tag{12}$$

where N_y^J follows a homogeneous Poisson process independent of the equilibrium price and the noise. It has a constant intensity λ_y which represents the expected number of jump that occur per unit time. $\{c_{y,i}\}$ is a jump size which is independent and identically distributed random variables with mean μ_{c_y} and variance $\sigma^2_{c_y}$. Following Fan and Wang [9] and Lee and Mykland [13], it is assumed that the jump size is also independent of P^*_y, u_y and N_y^J for $y = A, B$.

Under the transaction price process, the conditional expectation of $Z^{(\pm)}_{\ell, j_A j_B}$ with $\ell > 0$ is expressed as:

$$
\begin{aligned}
(13)\ \mathrm{E}[Z^{(\pm)}_{\ell, j_A j_B}] &= \mathrm{E}[(\int_{t_{j_A-1}}^{\bar{t}^{(+)}_{j_A}} \sigma^{(1)}(s)dW(s) + \Delta\bar{J}_{j_A} + u_A(\bar{t}^{(+)}_{j_A}) - u_A(t_{j_A-1})) \\
&\quad (\int_{\underline{t}^{(+)}_{j_B-1}}^{t_{j_B}} \sigma^{(2)}(s)dW(s) + \Delta\underline{J}_{j_B} + u_B(t_{j_B}) - u_B(\underline{t}^{(+)}_{j_B-1}))] \\
&= \mathrm{E}[\Delta\bar{J}_{j_A}\Delta\underline{J}_{j_B}] + \mathrm{E}[(u_A(\bar{t}^{(+)}_{j_A}) - u_A(t_{j_A-1}))(u_B(t_{j_B}) - u_B(\underline{t}^{(+)}_{j_B-1}))] \\
&= \mathrm{E}[\Delta\bar{J}_{j_A}]\mathrm{E}[\Delta\underline{J}_{j_B}] - \mathrm{E}[u_A(t_{j_A-1})u_B(t_{j_B})] \\
&= \lambda_A\lambda_B(\bar{t}^{(+)}_{j_A} - t_{j_A-1})(t_{j_B} - \underline{t}^{(+)}_{j_B-1})\mu_{c_A}\mu_{c_B} - \gamma(\ell),
\end{aligned}
$$

where $\Delta\bar{J}_{j_A} := J(\bar{t}^{(+)}_{j_A}) - J(t_{j_A-1})$ and $\Delta\underline{J}_{j_B} := J(t_{j_B}) - J(\underline{t}^{(+)}_{j_B-1})$. $\sigma^{(1)}$ and $\sigma^{(2)}$ are the first and second rows of $\mathbf{\Sigma}$ in (1). Let the maximal length of extended intervals for asset A and B be $\tilde{\Delta} := \max_{j_A, j_B}\{(\bar{t}^{(+)}_{j_A} - t_{j_A-1}),\ (t_{j_B} - \underline{t}^{(+)}_{j_B-1})\}$. Then, we obtain the following proposition for the bias of the noise cross and auto covariance estimator.

Proposition 3.1. *Suppose that the transaction price process is given in (11). The bias of the noise autocovariance $\hat{\gamma}_A(\ell)$ is negative. The sign of the bias of the noise cross covariance estimator $\hat{\gamma}(\ell)$ depends on the sign of $\mu_{c_A}\mu_{c_B}$ which is the product of the mean of jump size for asset A and B. Further, we have*

$$(14) \qquad | \mathrm{E}[\hat{\gamma}(\ell) - \gamma(\ell)] | \leq \lambda_A\lambda_B\tilde{\Delta}^2 | \mu_{c_A}\mu_{c_B} | = O(\tilde{\Delta}^2),$$

$$(15) \qquad | \mathrm{E}[\hat{\gamma}_A(\ell) - \gamma_A(\ell)] | \leq \lambda_A^2\tilde{\Delta}^2\mu_{c_A}^2 = O(\tilde{\Delta}^2),$$

The bias of the cross and auto covariance estimators are asymptotically negligible as the sampling frequency increases.

It is noted that $\tilde{\Delta}$ does not approach to zero as the sampling frequency increases since the noise covariance estimators (7) and (9) are based on the expanded intervals which do not converge to zero. Although the bias of the noise covariance estimator is not zero, the order of the bias is $O(\tilde{\Delta}^2)$ and the intensities λ_A, λ_B are quite small. Thus, the bias is asymptotically negligible. To see whether the influence of the rare jumps is negligible in finite sample, we conduct Monte Carlo simulation in Section 4.1.

3.2 Endogenous Noise and Noise Covariance Estimator

Ubukata and Oya's [16] covariance estimator is derived under the independence assumption between the equilibrium price and the noise. In some ways it is more natural to work with the noise which is correlated with equilibrium price for the same asset. Hence we need to see whether the noise covariance estimator works enough when the noise is endogenous. This section allows for the contemporaneous correlation between latent equilibrium return and noise:

$$\text{(16)} \qquad \mathrm{Cov}[(P_A^*(t_{i_A}) - P_A^*(t_{i_A-1})), u_A(t_{j_A})] = \begin{cases} \gamma_{e,i_A} & \text{if } i_A = j_A, \\ 0 & \text{otherwise.} \end{cases}$$

To make the analysis clear, we suppose that the transaction price process does not have the jump component. Under the transaction price process with (16), the conditional expectation of $Z^{(\pm)}_{\ell,i_A j_A}$ is expressed as:

- For $\ell = t_{j_A-1} - t_{i_A} = 0$:

$$\begin{aligned}
\text{(17) } \mathrm{E}[Z^{(\pm)}_{0,i_A j_A}] &= \mathrm{E}\Big[\Big(\int_{t_{j_A-1}}^{\bar{t}^{(+)}_{j_A}} \sigma^{(1)}(s)dW(s) + u_A(\bar{t}^{(+)}_{j_A}) - u_A(t_{j_A-1})\Big) \\
&\quad \Big(\int_{\underline{t}^{(-)}_{i_A-1}}^{t_{i_A}} \sigma^{(1)}(s)dW(s) + u_A(t_{i_A}) - u_A(\underline{t}^{(-)}_{i_A-1})\Big)\Big] \\
&= -\mathrm{E}\Big[\int_{\underline{t}^{(-)}_{i_A-1}}^{t_{i_A}} \sigma^{(1)}(s)dW(s) u_A(t_{j_A-1})\Big] - \mathrm{E}[u_A(t_{j_A-1})u_A(t_{i_A})] \\
&= -\mathrm{E}\Big[\Big\{\int_{t_{i_A-1}}^{t_{i_A}} \sigma^{(1)}(s)dW(s) + \int_{\underline{t}^{(-)}_{i_A-1}}^{t_{i_A-1}} \sigma^{(1)}(s)dW(s)\Big\} u_A(t_{j_A-1})\Big] \\
&\quad -\mathrm{E}[u_A(t_{j_A-1})u_A(t_{i_A})] \\
&= -\gamma_{e,i_A} - \gamma_A(0),
\end{aligned}$$

- For $\ell = t_{j_A-1} - t_{i_A} > 0$:

$$\text{(18)} \qquad \mathrm{E}[Z^{(\pm)}_{\ell,i_A j_A}] = -\gamma_A(\ell).$$

Let $\{\gamma_{e,k}\}_{k=1}^{N^{(\pm)}_{A,0}}$ be the set of γ_{e,i_A} for all i_A such that $i_A = j_A$. Then, we obtain the following proposition for the bias of the noise autocovariance estimator.

Proposition 3.2. *The bias of the noise autocovariance estimator $\hat{\gamma}_A(\ell)$ induced by endogenous noise is:*

$$\text{(19)} \qquad \mathrm{E}[\hat{\gamma}_A(\ell) - \gamma_A(\ell)] = \begin{cases} \frac{1}{N^{(\pm)}_{A,0}} \sum_{k=1}^{N^{(\pm)}_{A,0}} \gamma_{e,k} & \text{if } \ell = 0, \\ 0 & \text{if } \ell > 0. \end{cases}$$

The bias of the autocovariance estimator with $\ell > 0$ is zero under the correlation structure in (16) because the pair of nonoverlapping intervals with $\ell > 0$ is not adjacent. The order of $\frac{1}{N_{A,0}^{(\pm)}} \sum_{k=1}^{N_{A,0}^{(\pm)}} \gamma_{e,k}$ is $O(\tilde{\Delta}_A^{1/2})$ where $\tilde{\Delta}_A := \max_{i_A} \{(t_{i_A} - t_{i_A-1})\}$ is defined as the maximal length of observed intervals for asset A. Although Proposition 3.2 implies that the bias of noise autocovariance estimator with $\ell = 0$ is asymptotically negligible as the sampling frequency increases, the order is larger than case with the presence of jumps described in (14) and (15). The endogeneity of noise would be more influential to the noise covariance estimator than the jumps in finite sample. Thus, Section 4.2 will examine the impact of endogenous noise on the noise autocovariance estimator through Monte Carlo simulation.

4. Monte Carlo Simulation

As described in previous section, the bias of the noise covariance estimator depends on the intensity and size of the jump, the length of the extended interval and the endogeneity of the noise. We conduct Monte Carlo simulation to see the impact of them in finite sample.

4.1 The Impact of Rare Jumps on the Noise Covariance Estimator

The asymptotic argument in (14) and (15) of previous section requires that the sampling intervals approach to zero. The requirement cannot be perfectly met in real applications. In this subsection, we report simulation results which assess the finite sample performance of the noise covariance estimators in the presence of the jumps.

We simulate over the time interval $[0, T]$ which is thought to cover 6.5 hours ($T = 23400$ seconds) as with the NYSE and NASDAQ markets. This simulation exercise is based on regular and synchronous sampling with 10 second time interval. Two series of transaction prices, which represent the sum of the equilibrium price plus jumps and noise, are sampled synchronously at equidistant time.

It is assumed that the equilibrium price processes follow the stochastic differential equations introduced by Voev and Lunde [17]:

$$dP_y^*(t) = \sigma_y(t)\left[\sqrt{1-\xi_y^2}\, dW_y^{(1)}(t) + \xi_y\, dW_y^{(2)}(t)\right],$$
$$d\sigma_y^2(t) = \theta_y(\kappa_y - \sigma_y^2(t))\, dt + \phi_y \sigma_y^2(t)\, dW_y^{(2)}(t), \quad y = A, B,$$

where $W_y^{(\cdot)}$ represents a standard Brownian motion and $\sigma_y^2(t)$ follows the generalized autoregressive conditional heteroskedasticity (GARCH) diffusion process. When $W_A^{(1)}$ and $W_B^{(1)}$ are correlated with $d\langle W_A^{(1)}, W_B^{(1)}\rangle_t = \rho^*(t)dt$, the anti-Fisher transformation can be used to generate stochastic correlation $\rho^*(t)$. The values of parameters are the same as in Ubukata and Oya [16].

To generate the equilibrium price processes plus jumps, we choose the numbers of jump that occurs during the trading period as $\lambda_y T = 1$ to 5 with $y = A, B$.

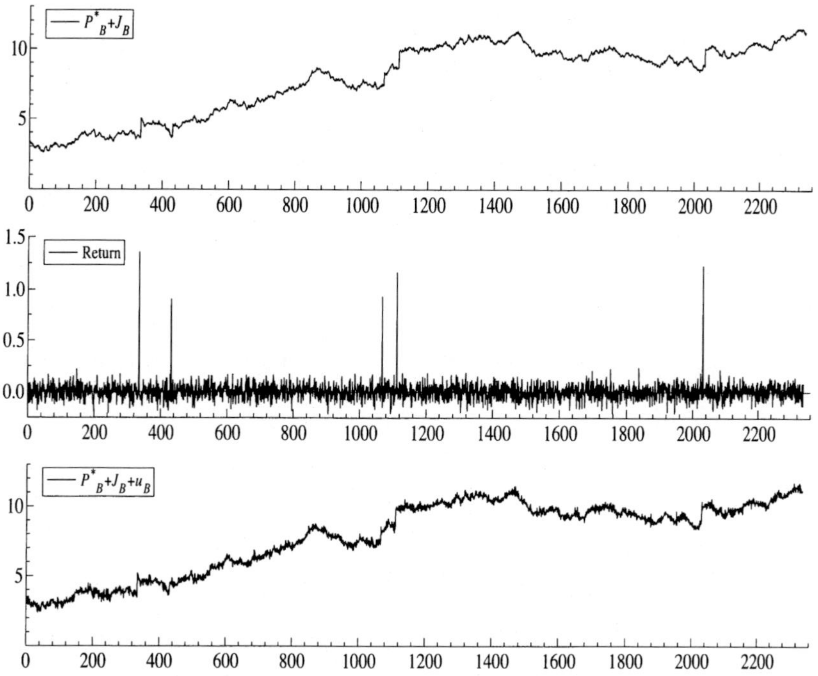

Figure 1. Simulated realizations in asset B, for 6.5 hours, using 10 second returns under the jump size parameters $(\lambda_B T, \mu_{c_B}, \sigma_{c_B}) = (5, 1, 3\overline{\sigma_B}(t))$.

The jump size c_y is generated by standard normal distribution with the mean of jump size being set as μ_y = -1, 0 and 1. Following Lee and Mykland [13], the standard deviations of jump sizes with three different magnitudes are assumed at $\sigma_{c_y} = 3\overline{\sigma_y}(t)$, $2\overline{\sigma_y}(t)$ and $\overline{\sigma_y}(t)$, where $\overline{\sigma_y}(t)$ is the mean of volatility $\sigma_y(t)$.

The microstructure noise is generated with the following bivariate MA(2) model every a transaction time:

$$\begin{bmatrix} u_A(t) \\ u_B(t) \end{bmatrix} = \begin{bmatrix} \epsilon_A(t) \\ \epsilon_B(t) \end{bmatrix} - \begin{bmatrix} 0.5 & 0.1 \\ 0.4 & 0.1 \end{bmatrix} \begin{bmatrix} \epsilon_A(t-1) \\ \epsilon_B(t-1) \end{bmatrix} + \begin{bmatrix} 0.4 & 0.15 \\ 0.15 & 0.3 \end{bmatrix} \begin{bmatrix} \epsilon_A(t-2) \\ \epsilon_B(t-2) \end{bmatrix},$$

where $(\epsilon_A(t), \epsilon_B(t))'$ is a bivariate Gaussian white noise. Thus, we can generate the transaction price process at regular and synchronous sampling interval:

$$P_A(t) = P_A^*(t) + J_A(t) + u_A(t),$$
$$P_B(t) = P_B^*(t) + J_B(t) + u_B(t).$$

Figure 1 plots the simulated realizations of length 6.5 hours at the 10 second frequency, with the jump size parameters $(\lambda_B T, \mu_{c_B}, \sigma_{c_B}) = (5, 1, 3\overline{\sigma_B}(t))$. We can see

Table 1. Bias and RMSE (percent) of noise cross and auto covariance estimators in case of no jumps.

	true covariances			bias×100			RMSE×100		
ℓ	$\gamma(\ell)$	$\gamma_A(\ell)$	$\gamma_B(\ell)$	$\hat{\gamma}(\ell)$	$\hat{\gamma}_A(\ell)$	$\hat{\gamma}_B(\ell)$	$\hat{\gamma}(\ell)$	$\hat{\gamma}_A(\ell)$	$\hat{\gamma}_B(\ell)$
−4	0			−0.003			0.213		
−3	0			0.024			0.232		
−2	0.009			−0.042			0.261		
−1	−0.018			−0.097			0.498		
0	0.027	0.044	0.022	−0.169	−0.147	−0.092	0.551	0.776	0.444
1	−0.015	−0.023	−0.011	−0.187	−0.110	−0.075	0.582	0.643	0.369
2	0.008	0.014	0.006	−0.071	−0.055	−0.027	0.274	0.332	0.208
3	0	0	0	0.047	0.025	0.016	0.230	0.286	0.191
4	0	0	0	−0.019	−0.006	−0.001	0.218	0.249	0.184

Note: In this simulation settings, the transaction price process follows the sum of the equilibrium price and noise without the rare jumps. The second, third and fourth columns represent true cross and auto covariances $\gamma(\ell)$, $\gamma_A(\ell)$ and $\gamma_B(\ell)$. In the regular and synchronous sampling, $\ell = 1$ represents one tick time whose length is equal to the sampling interval (10 seconds).

from the first and second panels that there are five large changes of prices. The transaction price process in the third panel looks similar to the plots of intraday prices in the financial econometrics literature.

Based on one thousand iterations, Table 1 represents a bias and root mean squared error (RMSE) of noise cross and auto covariance estimators $\hat{\gamma}(\ell)$, $\hat{\gamma}_A(\ell)$ and $\hat{\gamma}_B(\ell)$ under the transaction price process without jumps. The biases are quite small because the noise covariance estimator is unbiased and consistent in case of no jumps. The values of bias and RMSE are used as the experimental benchmark to compare the results in case with jumps.

Tables 2 and 3 contain the bias and RMSE of the noise covariance estimator in the presence of jumps. In each row, we can see to what extent the increase of number of jump $\lambda_y T$ affect to the bias and RMSE of the noise covariance estimator. The biases and RMSEs are almost the same as those in case of no jumps even in the frequent arrivals of jumps $\lambda_y T = 5$. Each column across all panels shows the effects by changing the standard deviation of jumps σ_{c_y} when the the number of jump $\lambda_y T$ and the mean of jump μ_{c_y} are fixed. The biases and RMSEs do not change as the variance of jumps σ_{c_y} increases. The reason for these results is clearly that the noise covariance terms dominate the bias term caused by jump component. This is consistent with the theoretical implication in Proposition 3.1. We omit the results in case of mean of jump size $\mu_y = 0, -1$ since the results for the bias and RMSE are almost the same as in Tables 2 and 3. Thus, the noise covariance estimators are not substantially affected by the presence of the jumps.

4.2 The Impact of Endogenous Noise on the Estimator

In this section we report the Monte Carlo results which assess the accuracy of the noise covariance estimator when the noise is endogenous. For simplicity, we

Table 2. Bias (percent) of cross and auto covariance estimators in the presence of jumps with the parameters $(\lambda_y T, \mu_{cy}, \sigma_{cy})$.

bias × 100

ℓ	(1, 1, 1$\overline{\sigma}(t)$) $\hat{\gamma}(\ell)$	$\hat{\gamma}_A(\ell)$	$\hat{\gamma}_B(\ell)$	(2, 1, 1$\overline{\sigma}(t)$) $\hat{\gamma}(\ell)$	$\hat{\gamma}_A(\ell)$	$\hat{\gamma}_B(\ell)$	(3, 1, 1$\overline{\sigma}(t)$) $\hat{\gamma}(\ell)$	$\hat{\gamma}_A(\ell)$	$\hat{\gamma}_B(\ell)$	(4, 1, 1$\overline{\sigma}(t)$) $\hat{\gamma}(\ell)$	$\hat{\gamma}_A(\ell)$	$\hat{\gamma}_B(\ell)$	(5, 1, 1$\overline{\sigma}(t)$) $\hat{\gamma}(\ell)$	$\hat{\gamma}_A(\ell)$	$\hat{\gamma}_B(\ell)$
-4	-0.004			-0.005			-0.011			-0.021			-0.012		
-3	0.019			0.024			0.022			0.020			0.025		
-2	-0.044			-0.045			-0.067			-0.055			-0.048		
-1	-0.113			-0.130			-0.128			-0.137			-0.143		
0	-0.177	-0.149	-0.096	-0.194	-0.137	-0.112	-0.204	-0.126	-0.136	-0.188	-0.157	-0.115	-0.191	-0.128	-0.115
1	-0.185	-0.114	-0.078	-0.191	-0.096	-0.093	-0.227	-0.096	-0.120	-0.190	-0.125	-0.097	-0.213	-0.117	-0.106
2	-0.075	-0.049	-0.028	-0.084	-0.053	-0.036	-0.086	-0.059	-0.049	-0.078	-0.072	-0.053	-0.074	-0.058	-0.039
3	0.051	0.024	0.017	0.048	0.016	0.022	0.043	0.016	0.005	0.038	0.024	0.021	0.034	0.024	0.022
4	-0.021	-0.002	0.000	-0.022	-0.007	-0.001	-0.029	-0.009	-0.021	-0.030	-0.018	-0.018	-0.026	-0.028	-0.009

ℓ	(1, 1, 2$\overline{\sigma}(t)$) $\hat{\gamma}(\ell)$	$\hat{\gamma}_A(\ell)$	$\hat{\gamma}_B(\ell)$	(2, 1, 2$\overline{\sigma}(t)$) $\hat{\gamma}(\ell)$	$\hat{\gamma}_A(\ell)$	$\hat{\gamma}_B(\ell)$	(3, 1, 2$\overline{\sigma}(t)$) $\hat{\gamma}(\ell)$	$\hat{\gamma}_A(\ell)$	$\hat{\gamma}_B(\ell)$	(4, 1, 2$\overline{\sigma}(t)$) $\hat{\gamma}(\ell)$	$\hat{\gamma}_A(\ell)$	$\hat{\gamma}_B(\ell)$	(5, 1, 2$\overline{\sigma}(t)$) $\hat{\gamma}(\ell)$	$\hat{\gamma}_A(\ell)$	$\hat{\gamma}_B(\ell)$
-4	-0.005			-0.006			-0.011			-0.022			-0.011		
-3	0.017			0.023			0.024			0.019			0.024		
-2	-0.045			-0.046			-0.067			-0.056			-0.048		
-1	-0.115			-0.128			-0.123			-0.138			-0.140		
0	-0.177	-0.149	-0.097	-0.193	-0.136	-0.111	-0.201	-0.125	-0.136	-0.189	-0.157	-0.113	-0.190	-0.128	-0.116
1	-0.184	-0.114	-0.079	-0.191	-0.094	-0.092	-0.227	-0.096	-0.120	-0.192	-0.125	-0.096	-0.212	-0.116	-0.107
2	-0.074	-0.049	-0.028	-0.084	-0.054	-0.036	-0.087	-0.058	-0.048	-0.077	-0.073	-0.051	-0.076	-0.059	-0.039
3	0.050	0.024	0.017	0.047	0.017	0.023	0.043	0.015	0.006	0.039	0.024	0.021	0.034	0.024	0.022
4	-0.021	-0.002	0.000	-0.022	-0.008	-0.001	-0.030	-0.009	-0.021	-0.030	-0.020	-0.017	-0.027	-0.029	-0.009

ℓ	(1, 1, 3$\overline{\sigma}(t)$) $\hat{\gamma}(\ell)$	$\hat{\gamma}_A(\ell)$	$\hat{\gamma}_B(\ell)$	(2, 1, 3$\overline{\sigma}(t)$) $\hat{\gamma}(\ell)$	$\hat{\gamma}_A(\ell)$	$\hat{\gamma}_B(\ell)$	(3, 1, 3$\overline{\sigma}(t)$) $\hat{\gamma}(\ell)$	$\hat{\gamma}_A(\ell)$	$\hat{\gamma}_B(\ell)$	(4, 1, 3$\overline{\sigma}(t)$) $\hat{\gamma}(\ell)$	$\hat{\gamma}_A(\ell)$	$\hat{\gamma}_B(\ell)$	(5, 1, 3$\overline{\sigma}(t)$) $\hat{\gamma}(\ell)$	$\hat{\gamma}_A(\ell)$	$\hat{\gamma}_B(\ell)$
-4	-0.004			-0.007			-0.010			-0.023			-0.011		
-3	0.017			0.023			0.024			0.020			0.023		
-2	-0.045			-0.046			-0.067			-0.057			-0.048		
-1	-0.113			-0.126			-0.125			-0.144			-0.145		
0	-0.177	-0.150	-0.098	-0.193	-0.136	-0.112	-0.202	-0.126	-0.135	-0.193	-0.159	-0.116	-0.194	-0.122	-0.119
1	-0.185	-0.115	-0.080	-0.193	-0.093	-0.092	-0.225	-0.098	-0.118	-0.194	-0.127	-0.099	-0.214	-0.110	-0.109
2	-0.073	-0.050	-0.028	-0.085	-0.054	-0.036	-0.087	-0.057	-0.049	-0.078	-0.068	-0.051	-0.079	-0.063	-0.041
3	0.049	0.025	0.016	0.047	0.017	0.023	0.045	0.016	0.006	0.039	0.020	0.020	0.034	0.026	0.022
4	-0.020	-0.003	-0.001	-0.022	-0.007	-0.001	-0.030	-0.010	-0.021	-0.029	-0.017	-0.017	-0.028	-0.030	-0.011

Note: In this simulation settings, the transaction price process follows the sum of the equilibrium price plus jumps and noise. The numbers of jump that occur during the trading period $\lambda_y T$ is set to be 1 to 5. The mean of jump size takes $\mu_y = 1$ and the variance of jump size is equal to 300 % to 100 % of $\overline{\sigma_y}(t)$ which is the mean of volatility $\sigma_y(t)$.

Table 3. RMSE (percent) of cross and auto covariance estimators in the presence of jumps with the parameters $(\lambda T, \mu_{c_y}, \sigma_{c_y})$.

RMSE × 100

	$(1,1,1\overline{\sigma}(t))$			$(2,1,1\overline{\sigma}(t))$			$(3,1,1\overline{\sigma}(t))$			$(4,1,1\overline{\sigma}(t))$			$(5,1,1\overline{\sigma}(t))$		
ℓ	$\hat\gamma(\ell)$	$\hat\gamma_A(\ell)$	$\hat\gamma_B(\ell)$	$\hat\gamma(\ell)$	$\hat\gamma_A(\ell)$	$\hat\gamma_B(\ell)$	$\hat\gamma(\ell)$	$\hat\gamma_A(\ell)$	$\hat\gamma_B(\ell)$	$\hat\gamma(\ell)$	$\hat\gamma_A(\ell)$	$\hat\gamma_B(\ell)$	$\hat\gamma(\ell)$	$\hat\gamma_A(\ell)$	$\hat\gamma_B(\ell)$
−4	0.221			0.218			0.219			0.240			0.247		
−3	0.237			0.254			0.243			0.264			0.267		
−2	0.276			0.270			0.269			0.278			0.299		
−1	0.513			0.524			0.535			0.538			0.554		
0	0.564	0.787	0.454	0.572	0.742	0.475	0.597	0.746	0.494	0.578	0.807	0.484	0.590	0.775	0.502
1	0.583	0.651	0.380	0.613	0.614	0.394	0.626	0.618	0.415	0.603	0.662	0.411	0.619	0.636	0.426
2	0.275	0.329	0.208	0.277	0.336	0.230	0.279	0.340	0.212	0.274	0.354	0.226	0.283	0.367	0.244
3	0.234	0.288	0.193	0.246	0.291	0.206	0.245	0.296	0.205	0.278	0.314	0.216	0.271	0.322	0.227
4	0.223	0.248	0.184	0.239	0.265	0.201	0.226	0.258	0.185	0.241	0.271	0.199	0.232	0.290	0.210

	$(1,1,2\overline{\sigma}(t))$			$(2,1,2\overline{\sigma}(t))$			$(3,1,2\overline{\sigma}(t))$			$(4,1,2\overline{\sigma}(t))$			$(5,1,2\overline{\sigma}(t))$		
ℓ	$\hat\gamma(\ell)$	$\hat\gamma_A(\ell)$	$\hat\gamma_B(\ell)$	$\hat\gamma(\ell)$	$\hat\gamma_A(\ell)$	$\hat\gamma_B(\ell)$	$\hat\gamma(\ell)$	$\hat\gamma_A(\ell)$	$\hat\gamma_B(\ell)$	$\hat\gamma(\ell)$	$\hat\gamma_A(\ell)$	$\hat\gamma_B(\ell)$	$\hat\gamma(\ell)$	$\hat\gamma_A(\ell)$	$\hat\gamma_B(\ell)$
−4	0.221			0.216			0.218			0.241			0.248		
−3	0.236			0.251			0.242			0.264			0.268		
−2	0.275			0.268			0.268			0.279			0.298		
−1	0.513			0.520			0.530			0.539			0.553		
0	0.564	0.787	0.454	0.569	0.742	0.475	0.596	0.747	0.493	0.580	0.806	0.481	0.591	0.777	0.504
1	0.580	0.651	0.381	0.612	0.612	0.393	0.626	0.618	0.415	0.606	0.662	0.409	0.621	0.638	0.429
2	0.275	0.329	0.208	0.276	0.335	0.231	0.280	0.338	0.212	0.274	0.354	0.226	0.282	0.367	0.246
3	0.234	0.288	0.194	0.246	0.292	0.205	0.245	0.297	0.206	0.278	0.313	0.217	0.272	0.326	0.229
4	0.223	0.248	0.184	0.239	0.265	0.201	0.227	0.259	0.185	0.242	0.271	0.197	0.233	0.292	0.212

	$(1,1,3\overline{\sigma}(t))$			$(2,1,3\overline{\sigma}(t))$			$(3,1,3\overline{\sigma}(t))$			$(4,1,3\overline{\sigma}(t))$			$(5,1,3\overline{\sigma}(t))$		
ℓ	$\hat\gamma(\ell)$	$\hat\gamma_A(\ell)$	$\hat\gamma_B(\ell)$	$\hat\gamma(\ell)$	$\hat\gamma_A(\ell)$	$\hat\gamma_B(\ell)$	$\hat\gamma(\ell)$	$\hat\gamma_A(\ell)$	$\hat\gamma_B(\ell)$	$\hat\gamma(\ell)$	$\hat\gamma_A(\ell)$	$\hat\gamma_B(\ell)$	$\hat\gamma(\ell)$	$\hat\gamma_A(\ell)$	$\hat\gamma_B(\ell)$
−4	0.221			0.216			0.219			0.243			0.251		
−3	0.236			0.252			0.242			0.267			0.269		
−2	0.275			0.268			0.269			0.281			0.298		
−1	0.512			0.518			0.532			0.546			0.555		
0	0.562	0.792	0.456	0.569	0.743	0.476	0.597	0.747	0.492	0.586	0.801	0.484	0.594	0.773	0.508
1	0.580	0.656	0.383	0.615	0.612	0.394	0.626	0.619	0.415	0.610	0.659	0.411	0.626	0.635	0.433
2	0.274	0.329	0.208	0.276	0.335	0.230	0.281	0.340	0.212	0.275	0.356	0.229	0.282	0.370	0.249
3	0.235	0.290	0.197	0.247	0.292	0.207	0.245	0.298	0.206	0.280	0.312	0.219	0.277	0.328	0.232
4	0.223	0.248	0.185	0.239	0.264	0.201	0.228	0.261	0.185	0.243	0.273	0.201	0.233	0.294	0.214

Note: In this simulation settings, the transaction price process follows the sum of the equilibrium price plus jumps and noise. The numbers of jump that occur during the trading period $\lambda_y T$ is set to be 1 to 5. The mean of jump size takes $\mu_y = 1$ and the variance of jump size is equal to 300 % to 100 % of $\overline{\sigma_y}(t)$ which is the mean of volatility $\sigma_y(t)$.

conduct a simulation experiment using a univariate framework. The equilibrium price process follows affine stochastic volatility model as in Heston [11]:

$$dP_A^*(t) = \sigma(t)dW(t)$$
$$d\sigma^2(t) = (0.3141 - 8.0369\sigma^2(t))dt + 0.1827\sigma(t)dB(t),$$

where W and B represent standard Brownian motions. The parameter estimates are obtained by Bakshi, Yu, and Ou-Yang [2] using the market index volatility approximated by the daily VIX. This simulation exercise is based on irregular and nonsynchronous sampling with 5 and 10 second time intervals. After equilibrium prices are randomly selected by Poisson sampling with mean durations of 5 and 10 seconds, we generate i.i.d. noise series of asset A $u_A(t)$ which is correlated with equilibrium returns $r^*_{j_A}$ by using the following simple linear model of endogeneity:

$$u_A(t_{j_A}) = \alpha r^*_{j_A} + \beta\epsilon(t_{j_A}), \tag{20}$$

where ϵ is random variable with standard normal distribution and $r^* \perp\!\!\!\perp \epsilon$. Given the variance of noise $\gamma_A(0)$ and equilibrium return-noise correlation coefficient ρ_e, the parameters α and β can be calculated as:

$$\alpha = \rho_e\sqrt{\gamma_A(0)/\mathrm{V}[r^*_{j_A}]}, \quad \beta = \sqrt{(1-\rho_e^2)\gamma_A(0)},$$

where the variance of equilibrium returns $\mathrm{V}[r^*_{j_A}]$ is estimated by the sample variance of equilibrium returns which are randomly selected by Poisson sampling. The variance of noise $\gamma_A(0)$ is determined by using noise-to-signal ratio defined as the noise variance divided by the integrated variance of equilibrium price process. The noise variance $\gamma_A(0)$ is set such that the noise-to-signal ratio for assets A amount to 0.005. Thus, the observed price process can be generated by the sum of the equilibrium prices and the noises.

Table 4 represents bias and RMSE of the noise variance estimator $\hat{\gamma}_A(\ell)$ with $\ell = 0$ and autocovariance estimator with $\ell = 5$ when the endogenous noise has six different magnitudes of the correlations ρ_e = -0.9 to 0.9 and there is exogenous noise $\rho_e = 0$, respectively. In this simulation settings, the true autocovariances $\gamma_A(\ell)$ with $\ell > 0$ are zero because the noise is independent of each other.

When $\rho_e = 0$, it is not surprising that the bias of the noise covariance estimator is virtually zero. On the other hand, we can see positively and negatively larger biases of $\hat{\gamma}_A(0)$ as ρ_e approaches to 0.9 and −0.9, respectively. While the bias of $\hat{\gamma}_A(5)$ does not change in all cases. These are consistent with Proposition 3.2. The final column in Table 4 represents an empirical size of test statistic for the significance of $\gamma_A(5)$. Although we see quite small upward bias of size, all empirical sizes are close to the nominal size of 5 percent. We confirm that the endogenous noise generated by (16) does not affect the noise autocovariance estimator $\hat{\gamma}_A(\ell)$ with $\ell > 0$.

Table 4. Bias and RMSE of noise autocovariance estimator under equilibrium price-noise correlation.

sampling interval: 10 seconds

	$\hat{\gamma}_A(0)$			$\hat{\gamma}_A(5)$		
ρ_e	bias	relative bias	RMSE	bias	RMSE	size
0.9	0.0077	0.203	0.0105	0.0005	0.0166	0.052
0.5	0.0043	0.113	0.0060	0.0001	0.0108	0.068
0.1	0.0009	0.022	0.0024	-0.0001	0.0087	0.061
0	0.0000	0.000	0.0022	-0.0002	0.0085	0.062
-0.1	-0.0009	-0.023	0.0024	-0.0002	0.0083	0.067
-0.5	-0.0045	-0.118	0.0060	-0.0002	0.0078	0.073
-0.9	-0.0083	-0.217	0.0106	0.0000	0.0091	0.046

sampling interval: 5 seconds

	$\hat{\gamma}_A(0)$			$\hat{\gamma}_A(5)$		
ρ_e	bias	relative bias	RMSE	bias	RMSE	size
0.9	0.0057	0.149	0.0076	0.0000	0.0048	0.049
0.5	0.0031	0.080	0.0043	0.0000	0.0030	0.041
0.1	0.0006	0.015	0.0017	0.0000	0.0027	0.062
0	-0.0001	-0.001	0.0016	0.0000	0.0026	0.067
-0.1	-0.0006	-0.017	0.0017	0.0000	0.0026	0.063
-0.5	-0.0032	-0.084	0.0043	0.0000	0.0024	0.053
-0.9	-0.0058	-0.152	0.0075	-0.0001	0.0033	0.036

Note: The market microstructure noises are generated by (20). ρ_e is the correlation coefficient between equilibrium price and noise. The last column is the empirical size of test statistic for the significance of $\gamma_A(5)$ (see Corollary 2 of Ubukata and Oya [16] for detail). When the sampling interval is equal to 10 seconds, $N_{A,\ell}^{(\pm)}$ for $\ell < 5$, which represents the sample size for asset A, is much smaller than $N_{2,\ell}^{(\pm)}$ for $\ell \geq 5$ because of Poisson sampling with the average duration being 10 seconds. Thus, we report the Monte Carlo results for $\hat{\gamma}_A(5)$ across sampling intervals.

In case of $\rho_e = -0.9$ and 0.9 with sampling interval being 10 seconds, the relative bias of $\hat{\gamma}_A(0)$ takes -0.217 and 0.203 which represent nonnegligible contamination by the endogenous noise. Although it seems that the noise covariance estimator is influenced by the higher equilibrium prices and noise correlations, its value would be not realistic. Hansen and Lunde's [10] empirical analysis reports empirical estimates of realized variance of equilibrium price, noise variance and equilibrium price-noise covariance for NYSE and NASDAQ stocks. From them the correlation coefficient of equilibrium price and noise approximately takes a range from -0.41 to 0.25 where the average is -0.09. When ρ_e = -0.1 and 0.1 and sampling interval is 10 seconds, the bias of $\hat{\gamma}_A(0)$ is almost same as that in case of ρ_e = 0. Even when ρ_e = -0.5 and 0.5, the bias and relative bias are not very large. Furthermore, the bias and RMSE of $\hat{\gamma}_A(0)$ with the higher trading intensity such that sampling intervals = 5 seconds are smaller than those with lower trading intensity (sampling intervals = 10 seconds). This results support that the noise covariance term dominates the equilibrium price-noise covariance term as the sampling intervals approach to zero. Figure 2 plots histograms of the relative

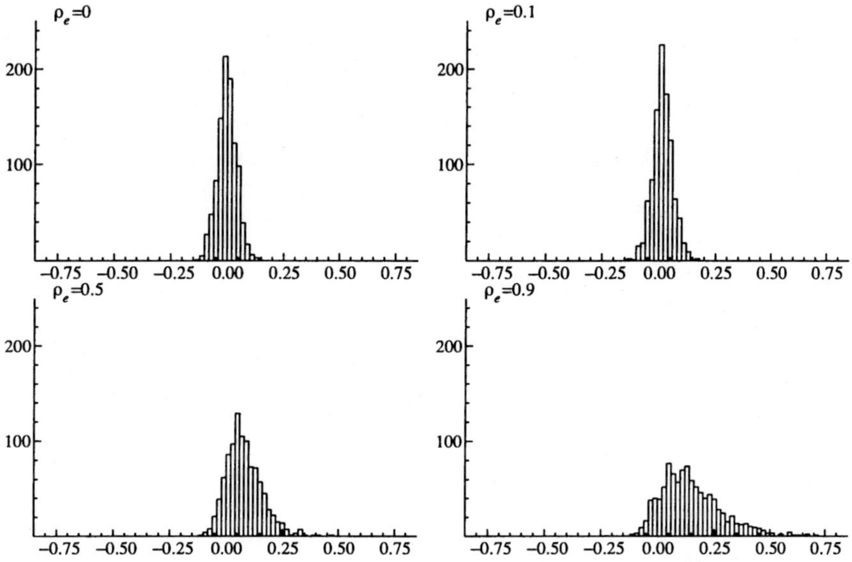

Figure 2. Histograms of the relative bias of $\hat{\gamma}_A(0)$ in case of ρ_e = 0, 0.1, 0.5 and 0.9.

bias of $\hat{\gamma}_A(0)$ in case of ρ_e = 0, 0.1, 0.5 and 0.9 where the sampling interval is equal to 5 seconds. We find that $\hat{\gamma}_A(0)$ has a high degree of concentration around the true $\gamma_A(0)$ except extremely high correlation between the equilibrium price and the noise.

5. Conclusion

This paper provides a detail analysis of the accuracy of noise covariance estimators when the rare jumps occur and there is an endogenous noise. Theoretical results show that biases of the covariance estimators are asymptotically negligible even in the presence of jump and endogenous noise. We also conduct Monte Carlo simulation to check the robustness of the noise covariance estimators. The simulation results are suggestive of the robustness of the estimator to the presence of jumps and endogenous noise while the empirically unrealistic high equilibrium price-noise correlation causes nonnegligible bias. Therefore, the independence assumption between the equilibrium price and noise does not statistically damage the analysis of asset prices with high trading intensities.

Acknowledgement

Ministry of Education, Science, Sports and Culture, Grant-in-Aid for Scientific Research (A), 90254135, 2008 and 18203013, 2008 and for Young Scientists (Start-up), 20830047, 2008.

References

1. Andersen, T. G., Bollerslev, T., Diebold, F. X., and Labys, P., The Distribution of Exchange Rate Volatility, *Journal of the American Statistical Association*, **96**, (2001), 42–55 [correction published in 2003, vol. 98, 501].
2. Bakshi, G. S., Yu, N., and Ou-Yang, H., Estimation of Continuous-time Models with an Application to Equity Volatility, *Journal of Financial Economics*, **82**, (2006), 227–249.
3. Bandi, F. M. and Russell, J. R., Full-information transaction costs, Working paper, Graduate School of Business, University of Chicago, (2006).
4. Bandi, F. M. and Russell, J. R., Microstructure Noise, Realized Variance, and Optimal Sampling, *Review of Economic Studies*, **75**, (2008), 339–369.
5. Barndorff-Nielsen, O. E. and Shephard, N., Power and Bipower Variation with Stochastic Volatility and Jumps, *Journal of Financial Econometrics*, **2**, (2005), 1–37.
6. Barndorff-Nielsen, O. E. and Shephard, N., Econometrics of Testing for Jumps in Financial Economics Using Bipower Variation, *Journal of Financial Econometrics*, **4**, (2005), 1–30.
7. Carlstein, E., The Use of Subseries Values for Estimating the Variance of a General Statistic from a Stationary Sequence, *The Annals of Statistics*, **14**, (1986), 1171–1179.
8. Eom, K. S., Ok, J. and Park, J.-H., Pre-trade transparency and market quality, *Journal of Financial Markets*, **10**, (2007), 319–341.
9. Fan, J. and Wang, Y., Multi-scale Jump and Volatility Analysis for High-Frequency Financial Data, *Journal of the American Statistical Association*, **102**, (2007), 1349–1362.
10. Hansen, P. R. and Lunde, A., Realized Variance and Market Microstructure Noise, *Journal of Business and Economic Statistics*, **24**, (2006), 127–161.
11. Heston, S., A Closed-Form Solution for Options with Stochastic Volatility with Applications to Bonds and Currency Options, *Review of Financial Studies*, **6**, (1993), 327–343.
12. Huang, X. and Tauchen, G., The Relative Contribution of Jumps to Total Price Variance, *Journal of Financial Econometrics*, **3**, (2005), 456–499.
13. Lee, S. S. and Mykland, P. A., Jumps in Financial Markets: A New Nonparametric Test and Jump Dynamics, *Review of Financial Studies*, to appear.
14. Meddahi, N., A Theoretical Comparison Between Integrated and Realized Volatility, *Journal of Applied Econometrics*, **17**, (2002), 479–508.
15. Mykland, P. A. and Zhang, L., Comment: A Selective Overview of Nonparametric Methods in Financial Econometrics, *Statistical Science*, **20**, (2005), 347–350.
16. Ubukata, M. and Oya, K., Estimation and Testing for Dependence in Market Microstructure Noise, *Journal of Financial Econometrics*, to appear.
17. Voev, V. and Lunde, A., Integrated Covariance Estimation using High-frequency Data in the Presence of Noise, *Journal of Financial Econometrics*, **5**, (2007), 68–104.
18. Zhang, L., Mykland, P. A., and Aït-Sahalia, Y., A tale of two time scales: Determining integrated volatility with noisy high frequency data, *Journal of the American Statistical Association*, **100**, (2005), 1394–1411.

Quanto Pre-washing for Jump Diffusion Models

Hoi Ying Wong* and Ka Yung Lau†

*Department of Statistics, The Chinese University of Hong Kong
E-mail: hywong@cuhk.edu.hk
†Department of Treasury Risk, Fubon Bank Hong Kong
E-mail : kayung.Lau@fubon.com

A quanto option is a derivatives security whose value depends on an exchange rate (domestic over foreign) and a foreign asset. Usually, the stochastic process of the exchange rate is given under the domestic risk-neutral measure whereas that of the foreign equity is defined under the foreign risk-neutral measure. As quanto options are quoted in the domestic currency world, the valuation requires the quanto pre-washing procedure which identifies the process of the foreign asset under the domestic risk-neutral measure and that of the domestic equivalent asset (the product of the exchange rate and the foreign asset value) under the domestic risk-neutral measure. This paper investigates the quanto pre-washing for the exchange rate and the foreign asset both following jump diffusion models with correlated Brownian motion component, common jumps and possibly dependent jump sizes. The use of the results is illustrated with the Merton jump diffusion model.
Key words: Quanto Pre-washing, Jump Diffusion, Option Pricing, Laplace Transform

1. Introduction

Quanto options are contingent claims whose payoff is determined by a financial price or index in one currency while the actual payoff is settled in a different currency. The payoffs of quanto options can be structured in a variety of combinations of foreign asset prices and exchange rates, thereby generating a rich set of investment and hedging opportunities. The wider classes of payoff structures allow investors to hedge a specific risk or pursue a particular speculation of international equity investment. Expositions on hedging properties of European quanto options can be found in Reiner (1992) and Toft and Reiner (1997) when the exchange rate and the foreign asset follow the Black-Scholes model. Under the Black-Scholes model, path-dependent quanto option pricing has been studied in Kwok and Wong (2000) and Dai, Wong and Kwok (2004).

However, it is well known that both the equity and exchange rate exhibit jumps in their values and the Black-Scholes model is far from adequate to model the empirical distribution of market data. Jump diffusion model is a widely accepted alternative to model stock price and exchange rate. It is a challenge to apply jump diffusion models to value quanto options because the jump diffusion market is incomplete in general and quanto options represent a class of options on multiple assets, or more specifically on two state variables, the exchange rate and the foreign equity. Thus, the difficulty arises from developing an appropriate link between risk-neutral processes of the domestic and the foreign markets. This establishment of such a link refers to the quanto pre-washing procedure in the literature (Reiner, 1992; Kwok and Wong, 2000).

Huang and Hung (2005) investigate the pricing of some specific quanto options with Levy processes. Our approach is significantly different from theirs. Although Huang and Hung (2005) consider general Levy processes, their method is only applicable to European quanto options that the payoff is linear with respect to either the exchange rate or the foreign asset or both. Their approach cannot be used to price fixed exchange rate quanto call/put, and quanto call/put on domestic equivalent asset, which is defined as the product of the exchange rate and the foreign asset. The major restriction of the Huang and Hung (2005) approach is that it does not explicitly give the corresponding process of the foreign asset under the domestic risk-neutral measure nor the process of the domestic equivalent asset under the domestic risk neutral measure. In addition, they do not take into account the possibility of dependent jump sizes.

This paper contributes to the literature by introducing an arbitrage-free based argument to formulate a quanto pre-washing framework to identify processes of the exchange rate and foreign asset under both the domestic and foreign risk neutral measure. Jump diffusion model is introduced by Merton (1976) to the financial industry. Merton's model assumes that jumps arrive according to a Poisson process with the log-jump-size being normally distributed. Kou (2002) extends the Merton model such that the jump size follows a double exponential distribution. To value quanto options, one has to specify the joint dynamics of the exchange rate and the foreign asset in the sense of jump diffusion models. Although our framework works for a general jump diffusion model, we concentrate on the Merton model with common jumps and jump sizes being bivariate normally distributed.

The remaining part of the paper is organized as follows. Section 2 introduces various types of quanto options in the financial market and jump diffusion models. Section 3 which contains the main result of this paper establishes the quanto prewashing procedure for quanto option pricing under general jump diffusion models. Section 4 demonstrates the use of the result with the Merton jump diffusion model. Conclusion is made in Section 5.

2. Quanto Options

Quanto options are claims contingent on an foreign asset, S, and are settled in a domestic currency world using the domestic-to-foreign exchange rate F. We now give several examples of quanto options.

1. Floating foreign exchange rate (FX) quanto call

$$\text{payoff: } F_T \max(S_T - K_f, 0), \tag{1}$$

where K_f is the strike price in foreign currency. This option is equivalent to a foreign call option translated to the domestic currency world using F_T on the maturity date T. The process of S_t in the foreign risk neutral measure, $\mathbb{F}$, suffices for the valuation.

2. Fixed FX quanto call

$$\text{payoff: } F_0 \max(S_T - K_f, 0), \tag{2}$$

where F_0 is a predetermined fixed exchange rate and K_f is the strike price in foreign currency. The valuation of this option requires the dynamics of S_t under the domestic risk neutral measure, $\mathbb{D}$.

3. Domestic equivalent equity quanto call

$$\text{payoff: } \max(F_T S_T - K_d, 0), \tag{3}$$

where K_d is the strike price in domestic currency. The valuation of this option requires the dynamics of $F_t S_t$ under $\mathbb{D}$.

In fact, European-style quanto options' payoff can generally be expressed as the form: $\Psi(S_T, F_T)$. Thus, the representation for the present value of such a quanto option in the domestic currency world is given by

$$e^{-r_d(T-t)} E^{\mathbf{D}} \left[\Psi(S_T, F_T) | \mathcal{F}_t \right], \tag{4}$$

where r_d is the constant interest rate in the domestic currency world and $\mathcal{F}_t$ is the filtration or information accumulated up to the present time t.

2.1 The Jump Diffusion Model

The jump diffusion model consists of two components: the diffusion part and the jump part. Consider the dynamic of F under the domestic risk neutral measure:

$$\frac{dF}{F} = [r_d - r_f - (k_F - 1)\lambda_F]\, dt + \sigma_F\, dW_F + X\, dN_F, \tag{5}$$

where $k_F = E^{\mathbf{D}}[e^x]$ with $X = e^x - 1$ being the jump size, r_d is the domestic constant interest rate, r_f is the foreign constant interest rate, σ_F is the diffusion volatility of the exchange rate, W_F is the Wiener process defined under $\mathbb{D}$ and N_F is the Poisson process with intensity λ_F. We assume that W_F, X and N_F are independent random variables for any fixed future time point.

The foreign asset is, however, defined under the foreign risk neutral measure. Specifically, The general dynamic of S is given by,

$$\frac{dS}{S} = [r_f - (k_S - 1)\lambda_S]\,dt + \sigma_S\, dW_S + Y\,dN_S, \tag{6}$$

where $k_S = E^{\mathbb{F}}[e^y]$ with $Y = e^y - 1$, σ_S is the constant diffusion volatility of the foreign asset return, W_S is the Wiener process under the foreign risk neutral measure $\mathbb{F}$ and N_S is the Poisson process with constant intensity λ_S. Similar to the case of the FX, we assume that W_S, N_S and Y are independent random variables for any fixed future point.

Certainly, the FX and the foreign asset could not be independent in general. The correlation structure between them can be characterised using the foreign measure $\mathbb{F}$. Specifically,

$$\begin{aligned} E^{\mathbb{F}}[dW_F\, dW_S] &= \rho\, dt, \\ E^{\mathbb{F}}[dN_F\, dN_S] &= \lambda\, dt, \\ \text{p.d.f. between two jump sizes} &= f(x, y), \end{aligned} \tag{7}$$

where p.d.f. stands for the joint probability density function.

To further illustrate the meaning of (7), let us consider the Merton (1976) model in which the jump sizes $x = \log(1 + X)$ and $y = \log(1 + Y)$ are normally distributed so that it is natural to assume that (x, y) follows a bivariate normal distribution. In such a situation, $f(x, y)$ is the density function of a bivariate normal distribution. There are many possible ways such that the second condition in (7) holds. For instance, $N_S = N_1 + N$ and $N_F = N_2 + N$, where N, N_1 and N_2 are independent Poisson processes with intensities λ, λ_1 and λ_2, respectively, such that $\lambda_S = \lambda_1 + \lambda$ and $\lambda_F = \lambda_2 + \lambda$. The first condition of (7) is clear.

3. Quanto Pre-washing

Prior to the pricing of quanto options, it is crucial for us to pre-wash the dynamics of F and S under both $\mathbb{D}$ and $\mathbb{F}$. It is useful to point out that $1/F$ can be traded in the foreign currency world so that both S and $1/F$ can be viewed as foreign assets. Thus, it is possible to have a derivative security which payoff depends on both S and $1/F$. This payoff function can be denoted as $\Psi(S, F)$. The building block for the quanto pre-washing is the following arbitrage-free condition on trading a special type of quanto options between the domestic and foreign currency worlds.

Theorem 3.1.

$$e^{-r_d(T-t)} E^{\mathbb{D}}\left[F_T \Psi(S_T, F_T) | \mathcal{F}_t \right] = e^{-r_f(T-t)} F_t \cdot E^{\mathbb{F}}\left[\Psi(S_T, F_T) | \mathcal{F}_t \right], \tag{8}$$

for any quanto options whose payoffs take the form $F_T\Psi(S_T, F_T)$ in the domestic currency world, where $\Psi(S, F)$ is a square-integrable deterministic function.

Proof. Consider a foreign derivative security whose payoff is $\Psi(S_T, F_T)$. In the foreign currency world, this derivative security has the fair price of

$$e^{-r_f(T-t)}E_t^{\mathbb{F}}[\Psi(S_T, F_T)].$$

Hence, a domestic investor can purchase this foreign security with the domestic price:

$$F_t \times \left\{e^{-r_f(T-t)}E_t^{\mathbb{F}}[\Psi(S_T, F_T)]\right\}.$$

Suppose that the domestic currency world offers a quanto option whose payoff is $F_T\Psi(S_T, F_T)$. Using the domestic measure, its present value is given by

$$e^{-r_d(T-t)}E_t^{\mathbb{D}}[F_T\Psi(S_T, F_T)].$$

In addition, these two derivative securities provide exactly the same terminal cash flow for the domestic investor, where the terminal cash flow is $F_T\Psi(S_T, F_T)$. By no-arbitrage argument, these two securities should have the same present value in the domestic currency world; otherwise, a simple "buy low sell high" strategy makes a risk-free profit. The proof is hence completed.

□

The value of the floating FX quanto call, which payoff is defined in (1), can be immediately obtained from Theorem 3.1. Specifically, its value is given by

$$F_t\left[e^{-r_f(T-t)}E^{\mathbb{F}}[\max(S_T - K_f, 0)]\right].$$

Neglecting the current FX rate, F_t, the remaining part is the European call price in the foreign currency world. Therefore, the floating FX quanto call price is ready.

It will be seen shortly that Theorem 3.1 is also very useful in deriving the processes of S and F under risk neutral measures of the two currency worlds. We begin with the FX rate which process is defined under the domestic measure in (5), from which it is easy to show that

$$E_t^{\mathbb{D}}[F_T^{\alpha}] = F_t^{\alpha}\exp[G^D(\alpha)(T-t)], \tag{9}$$

$$G^D(\alpha) := \mu_F^D\alpha + \frac{\sigma_F^2}{2}\alpha^2 + \lambda_F E^{\mathbb{D}}[e^{\alpha x} - 1], \tag{10}$$

where $\mu_F^D = r_d - r_f - \frac{\sigma_F^2}{2} - (k_F - 1)\lambda_F$. The function in (9) refers to the moment generating function (MGF) of the log-FX rate under the domestic measure and that in (10) is the corresponding Laplace exponent.

Consider a foreign derivative on $1/F$ with the payoff being $(1/F_T)^{-\alpha} = F_T^{\alpha}$. By Theorem 3.1, it is clear that

$$e^{-r_d(T-t)} E_t^{\mathbf{D}} \left[F_T^{\alpha+1} \right] = e^{-r_f(T-t)} F_t \cdot E_t^{\mathbf{F}} \left[F_T^{\alpha} \right],$$

which together with (9) and (10) imply the following result.

Proposition 3.1. *For any real constant α,*

$$E_t^{\mathbf{F}} \left[F_T^{\alpha} \right] = F_t^{\alpha} \exp[G^F(\alpha)(T-t)],, \tag{11}$$

where

$$\begin{aligned} G^F(\alpha) &= G^D(\alpha+1) - G^D(1) \\ &= (\mu_F^D + \sigma_F^2)\alpha + \frac{\sigma_F^2}{2}\alpha^2 + (k_F \lambda_F)\left(\frac{E^{\mathbf{D}}[e^{(\alpha+1)x}]}{E^{\mathbf{D}}[e^x]} - 1 \right). \end{aligned} \tag{12}$$

Proposition 3.1 links the foreign dynamics of the FX rate to its domestic dynamics because (11) gives the MGF for the log-FX rate under the foreign measure. It can be seen from (11) and (12) that the FX rate still follows a jump-diffusion under the foreign measure in which the drift and diffusion volatility remain the same but the jump intensity is changed to $k_F \lambda_F$ and the jump size is described by the MGF: $E^{\mathbf{F}}[e^{\alpha x}] = E^{\mathbf{D}}[e^{(\alpha+1)x}]/E^{\mathbf{D}}[e^x]$.

Recall that the foreign process of S has been defined in (6), and Proposition 3.1 obtains the foreign process for F. Therefore, it is possible to derive the joint MGF for the FX rate and the foreign asset under the foreign measure. Using (6), (7) and Proposition 3.1, it is easy to compute the joint MGF as follows.

Proposition 3.2. *For any real constants α and β,*

$$E_t^{\mathbf{F}}[F_T^{\alpha} S_T^{\beta}] = F_t^{\alpha} S_t^{\beta} \exp[G^F(\alpha,\beta)(T-t)], \tag{13}$$

where

$$\begin{aligned} G^F(\alpha,\beta) = {} & (\mu_F^D + \sigma_F^2)\alpha + \mu_S^F \beta + \frac{1}{2}(\sigma_F^2 \alpha^2 + 2\rho \sigma_F \sigma_S \alpha\beta + \sigma_S^2 \beta^2) \\ & + (k_F \lambda_F - \lambda) E^{\mathbf{D}}[e^{\alpha x} - 1] + (\lambda_S - \lambda) E^{\mathbf{F}}[e^{\beta y} - 1] \\ & + \lambda E^{\mathbf{F}}[e^{\alpha x + \beta y} - 1]. \end{aligned} \tag{14}$$

and $\mu_S^F = r_F - \frac{1}{2}\sigma_S^2 - (k_S - 1)\lambda_S$ is the drift of the log-value of S under the foreign measure.

As quanto options are traded in the domestic currency world, it is more important to have the joint MGF under the domestic measure, i.e. $E_t^{\mathbf{D}}[F^{\alpha} S^{\beta}]$. The domestic joint MGF can be derived from Theorem 3.1, in which $\Psi(S_T, F_T) = F_T^{\alpha-1} S_T^{\beta}$, and Proposition 3.2. The consequence is that

$$E_t^{\mathbf{D}}[F_T^{\alpha} S_T^{\beta}] = e^{(r_d - r_f)(T-t)} F_t E_t^{\mathbf{F}}[F_T^{\alpha-1} S_T^{\beta}].$$

Simple calculation shows the following.

Proposition 3.3. *For any real constants α and β,*

$$E_t^{\mathbf{D}}[F_T^{\alpha} S_T^{\beta}] = F_t^{\alpha} S_t^{\beta} \exp[G^D(\alpha,\beta)(T-t)], \tag{15}$$

where

$$G^D(\alpha,\beta) = G^F(\alpha-1,\beta) - G^F(-1,0), \tag{16}$$

and $G^F(\alpha,\beta)$ has been given in Proposition 3.2.

Using Proposition 3.2, an explicit expression for $G^D(\alpha,\beta)$ can be obtained as follows.

$$\begin{aligned} G^D(\alpha,\beta) &= \mu_F^D \alpha + (\mu_S^F - \rho\sigma_F\sigma_S)\beta + \frac{1}{2}(\sigma_F^2\alpha^2 + 2\rho\sigma_F\sigma_S\alpha\beta + \sigma_S^2\beta^2) \\ &\quad + (\lambda_F - k_F^{-1}\lambda)E^{\mathbf{D}}[e^{\alpha x} - 1] + (\lambda_S - \lambda)E^{\mathbf{F}}[e^{\beta y} - 1] \\ &\quad + k_F^{-1}\lambda\left[E^{\mathbf{F}}[e^{(\alpha-1)x+\beta y}]E^{\mathbf{D}}[e^{x}] - 1\right]. \end{aligned} \tag{17}$$

To value the fixed FX quanto call, whose payoff is given in (2), we need the process of S under the domestic risk neutral measure, $\mathbb{D}$. By substituting $\alpha = 0$, the MGF of $\log S$ under $\mathbb{D}$ can be obtained from Proposition 3.3 with β being the independent variable of the MGF. Whereas, the option value of the domestic equivalent equity call with payoff defined in (3) requires the MGF for the domestic equivalent asset, $F_T S_T$. This latter MGF can be also obtained from Proposition 3.3 by setting $\alpha = \beta$.

4. The Merton Jump Diffusion Model

Based on the process of the FX rate under $\mathbb{D}$ defined in (5), the Merton jump diffusion model further specifies that the log-jump-size $x = \log(1 + X)$ follows a normal distribution. Then, the dynamics of the log FX rate under $\mathbb{D}$ can be obtained from the Itô lemma, and that under $\mathbb{F}$ can be inferred from Proposition 3.1. By matching coefficients of the MGF in Proposition 3.1 and those from the Merton model, we arrive at the following pair of dynamics of F:

$$\begin{aligned} \mathbb{D} : d\log F &= (r_d - r_f - \sigma_F^2/2 - (k_F - 1)\lambda_F)\,dt + \sigma_F\,dW_F + x\,dN_F, \\ \mathbb{F} : d\log F &= (r_d - r_f + \sigma_F^2/2 - (k_F - 1)\lambda_F)\,dt + \sigma_F\,d\widehat{W}_F + \widehat{x}\,d\widehat{N}_F, \end{aligned} \tag{18}$$

where $\widehat{W}_F$ is the Wiener process under the measure $\mathbb{F}$, $\widehat{N}_F$ is the Poisson process under the measure $\mathbb{F}$, the intensities of N_F and $\widehat{N}_F$ are respectively λ_F and $k_F\lambda_F$, $x \sim \mathcal{N}(\mu_x, \sigma_x^2)$, and $\widehat{x} \sim \mathcal{N}(\mu_x + \sigma_x^2, \sigma_x^2)$.

The process of S is defined in (6) under $\mathbb{F}$. Applying Itô's lemma, we have

$$\mathbb{F} : d\log S = [r_f - \frac{\sigma_S^2}{2} - (k_S - 1)\lambda_S]\,dt + \sigma_S\, dW_S + y\,dN_S. \tag{19}$$

To complete the Merton model, it remains to specify the joint dynamic of S and F under $\mathbb{F}$. We model the jump sizes as a bivariate normal distribution with a correlation coefficient ρ_j. Therefore, the joint MGF of the pair of jump sizes $(\widehat{x}, y)$ under foreign measure is given by:

$$\exp\left[(\mu_x + \sigma_x^2)\alpha + \mu_y\beta + \frac{1}{2}(\sigma_x^2\alpha^2 + 2\rho_j\sigma_x\sigma_y\alpha\beta + \sigma_y^2\beta^2)\right].$$

In addition, we allow the two Wiener processes W_S and $\widehat{W}_F$ to be correlated with correlation coefficient ρ and the two Poisson processes to be correlated as well. Thus, $N_S = N_1 + N$ and $\widehat{N}_F = N_2 + N$ where N, N_1 and N_2 are independent Poisson processes under $\mathbb{F}$ with intensities λ, $\lambda_S - \lambda$ and $\lambda_F - \lambda$, respectively.

By substituting everything into (17), it is easy to obtain:

$$\begin{aligned} G^D(\alpha,\beta) = {} & \mu_F^D\alpha + (\mu_S^F - \rho\sigma_F\sigma_S)\beta + \frac{1}{2}(\sigma_F^2\alpha^2 + 2\rho\sigma_F\sigma_S\alpha\beta + \sigma_S^2\beta^2) \\ & + (\lambda_F - k_F^{-1}\lambda)[e^{\mu_x\alpha + \sigma_x^2\alpha^2/2} - 1] + (\lambda_S - \lambda)[e^{\mu_y\beta + \sigma_y^2\beta^2/2} - 1] \\ & + k_F^{-1}\lambda[e^{\mu_x\alpha + (\mu_y - \rho_j\sigma_x\sigma_y)\beta + \frac{1}{2}(\sigma_x^2\alpha^2 + 2\rho_j\sigma_x\sigma_y\alpha\beta + \sigma_y^2\beta^2)} - 1], \end{aligned} \tag{20}$$

where $k_F^{-1} = e^{-\mu_x - \sigma_x^2/2}$.

4.1 Closed Form Solutions

Consider the fixed FX quanto call: $F_0 \max(S_T - K_f, 0)$. The value of this option can be expressed as

$$F_0\left[e^{-r_d(T-t)}E_t^{\mathbf{D}}\left[\max(S_T - K_f, 0)\right]\right].$$

The computation of the expectation above requires the process of S under $\mathbb{D}$. Using (20) and setting $\alpha = 0$, we obtain the MGF for S under $\mathbb{D}$ through comparing coefficients of $G^D(0,\beta)$ and the MGF of the log-asset value under the Merton jump diffusion model, where β is the independent variable of the MGF. The result is given as follows.

$$\mathbb{D} : d\log S = (\mu_S^F - \rho\sigma_F\sigma_S)\,dt + \sigma_S\, dW_S + y\,dN_S + (y - \rho_j\sigma_x\sigma_y)\,dN, \tag{21}$$

where

$$\begin{aligned} E^{\mathbf{D}}[dN_S] &= \widehat{\lambda}_S\, dt = (\lambda_S - \lambda)dt, \\ E^{\mathbf{D}}[dN] &= \widehat{\lambda}\, dt = k_F^{-1}\lambda\, dt, \\ (x, y)' &\sim N\left(\begin{pmatrix}\mu_x \\ \mu_y\end{pmatrix}, \begin{pmatrix}\sigma_x^2 & \rho_j\sigma_x\sigma_y \\ \rho_j\sigma_x\sigma_y & \sigma_y^2\end{pmatrix}\right). \end{aligned}$$

Thus, the fixed FX quanto call option value becomes:

$$
\begin{aligned}
&F_0 e^{-r_d(T-t)} E_t^{\mathbf{D}}[(S_T - K_f)^+] \\
&= F_0 e^{-r_d(T-t)} E_t^{\mathbf{D}}\left[E^{\mathbf{D}}[(S_T - K_f)^+ | N = n, N_S = m]\right] \\
&= F_0 e^{-r_d(T-t)} \sum_{n,m=0}^{\infty} \left[S^{(n,m)}\Phi(d_1^{(n,m)}) - K_f \Phi(d_2^{(n,m)})\right] \mathbb{D}_t\,[N = n, N_S = m] \\
&= F_0 e^{-r_d(T-t)} \sum_{n,m=0}^{\infty} [S^{(n,m)}\Phi(d_1^{(n,m)}) - K_f \Phi(d_2^{(n,m)})] \\
&\quad \times \frac{e^{-(\widehat{\lambda}+\widehat{\lambda}_S)(T-t)}[\widehat{\lambda}(T-t)]^n(\widehat{\lambda}_S(T-t))^m}{n!m!}
\end{aligned}
$$

where $\Phi(\cdot)$ is the cumulative distribution function of a standard normal random variable,

$$
\begin{aligned}
S^{(n,m)} &= E_t^{\mathbf{D}}[S_T | N = n, N_S = m] = S_t e^{(\mu_S^F - \rho\sigma_F\sigma_S)(T-t) + n(\mu_y - \rho_j\sigma_x\sigma_y) + m\mu_y} \\
d_2^{(n,m)} &= \frac{\log(S^{(n,m)}/K_f) - 1/2[\sigma_S^2(T-t) + (n+m)\sigma_y^2]}{\sqrt{\sigma_S^2(T-t) + (n+m)\sigma_y^2}} \\
d_1^{(n,m)} &= d_2^{(n,m)} + \sqrt{\sigma_S^2(T-t) + (n+m)\sigma_y^2}.
\end{aligned}
$$

Alternatively, similar method can be applied to the valuation of the domestic equivalent equity call as well. By putting $\alpha = \beta = 1$ in (20), the domestic process of $S_t F_t$ is obtained as

$$
\mathbb{D} : d\log(SF) = \mu\, dt + \sigma\, dW + x\, dN_F + y\, dN_S + (x + y - \rho_j\sigma_x\sigma_y)\, dN
$$

Hence, the domestic equivalent equity call option value reads:

$$
\begin{aligned}
&e^{-r_d(T-t)} E_t^{\mathbf{D}}[(F_T S_T - K_d)^+] \\
&= e^{-r_d(T-t)} E_t^{\mathbf{D}}\left[E_t^{\mathbf{D}}[(F_T S_T - K_d)^+ | N, N_S, N_F]\right] \\
&= e^{-r_d(T-t)} \sum_{n,m,\ell=0}^{\infty} \left[\widehat{S}^{(n,m,\ell)}\Phi(d_1^{(n,m,\ell)}) - K_d \Phi(d_2^{(n,m,\ell)})\right] \mathbb{D}_t\,[N = n, N_S = m, N_F = \ell] \\
&= e^{-r_d(T-t)} \sum_{n,m,\ell=0}^{\infty} [\widehat{S}^{(n,m,\ell)}\Phi(d_1^{(n,m,\ell)}) - K_d \Phi(d_2^{(n,m,\ell)})] \\
&\quad \times \frac{e^{(\widehat{\lambda}+\widehat{\lambda}_S+\widehat{\lambda}_F)(T-t)}(\widehat{\lambda}(T-t))^n(\widehat{\lambda}_S(T-t))^m(\widehat{\lambda}_F(T-t))^\ell}{n!m!\ell!}
\end{aligned}
$$

where

$$\widehat{S}^{(n,m,\ell)} = E^{\mathbf{D}}[F_T S_T | N = n, N_S = m, N_F = \ell]$$
$$= F_t S_t e^{\mu(T-t)+n(\mu_x+\mu_y-\rho_j\sigma_x\sigma_y)+m\mu_y+\ell\mu_x}$$
$$d_2^{(n,m,\ell)} = \frac{\log(\hat{S}^{(n,m,\ell)}/K)}{\sqrt{\sigma^2(T-t)+(n+m)\sigma_y^2+\ell\sigma_x^2}}$$
$$d_1^{(n,m,\ell)} = d_2^{(n,m,\ell)} + \sqrt{\sigma^2(T-t)+(n+m)\sigma_y^2+\ell\sigma_x^2}.$$

4.2 Laplace Transform Solutions

We have demonstrated the use of the quanto pre-washing in pricing quanto options under the Merton jump diffusion model. The pricing formulas are derived in explicit form. However, it may not be possible for other jump diffusion models or Levy processes to obtain closed form solutions in the same manner. The major reason is that it is sometimes inconvenient for us to identify the precise stochastic processes for the FX rate and the foreign asset.

For European quanto option pricing, it, however, suffices to have the joint MGF of the FX rate and the foreign asset if the purpose is to determine the numerical values of the options. The key is the use of Laplace transform. Many European quanto option prices are linked to the MGF after taking Laplace transform on the pricing formula with respect to the log strike price.

Consider the fixed FX quanto call as an example. If the MGF of log foreign asset value is known under the domestic measure, then the following calculation can be performed. Let $k = -\log K_f$ where K_f is the strike price. Consider the Laplace transform on the pricing representation:

$$\mathcal{L}_{k,\alpha}\{C_X(S,K_f)\} = F_0 e^{-r_d(T-t)} \int_{-\infty}^{\infty} e^{-\beta k} E^{\mathbf{D}}[\max(S_T - e^{-k}, 0)]dk$$
$$= F_0 e^{-r_d(T-t)} \int_{-\infty}^{\infty} e^{-\beta k} \int_{-\infty}^{\infty} \max(e^Z - e^{-k}, 0) q(Z) dZ dk$$
$$= F_0 e^{-r_d(T-t)} \int_{-\infty}^{\infty} \int_{-Z}^{\infty} e^{-\beta k}(e^Z - e^{-k}) q(Z) dk dZ$$
$$(22) \qquad = F_0 e^{-r_d(T-t)} \int_{-\infty}^{\infty} \left(\frac{1}{\beta} e^{(\beta+1)Z} - \frac{1}{1+\beta} e^{(\beta+1)Z}\right) q(Z) dZ,$$

where $C_X(S,K_f)$ denotes the present value of the fixed FX rate quanto call option, $Z = \log S_T$, $q(Z)$ is the probability density function of Z under the domestic measure and $\beta > 0$. By the definition of MGF, we have

$$(23) \qquad \mathcal{L}_{k,\alpha}\{C_X(S,K_f)\} = F_0 e^{-r_d(T-t)} \left[\frac{1}{\beta} - \frac{1}{1+\beta}\right] \mathrm{MGF}(\beta+1),$$

$$(24) \qquad \mathrm{MGF}(\beta) = E_t^{\mathbf{D}}[S_T^{\beta}],$$

where the explicit solution of the MGF is given in Proposition 3.3 for a general jump diffusion model by setting $\alpha = 0$ and then performing Laplace inversion with respect to β.

Similarly, the domestic equivalent equity call can be numerically valued using Laplace transform. Specifically, it is easy to show with exactly the same procedure that

$$\mathcal{L}_{k,\alpha}\{C_D(S, K_f)\} = e^{-r_d(T-t)}\left[\frac{1}{\beta} - \frac{1}{1+\beta}\right]\text{MGF}^*(\beta+1), \tag{25}$$

$$\text{MGF}^*(\beta) = E_t^{\text{D}}[(F_T S_T)^{\beta}], \tag{26}$$

where $C_D(S, K_f)$ denotes the present value of the domestic equivalent equity call, and $\text{MGF}^*(\beta)$ can be obtained by setting $\alpha = \beta$ in Proposition 3.3.

There is a number of ways to do the Laplace inversion. Some efficient algorithms can be found in Wong and Lau (2008) with application in exotic option pricing.

5. Conclusion

This paper develops a unified quanto pre-washing framework for pricing quanto options under jump diffusion models, which have correlated diffusion component, correlated Poisson arrival jumps and correlated jump sizes between the foreign asset and the exchange rate. The key is the use of arbitrage-free argument and the proper use of the moment generating functions of the log-asset value, log FX rate and log equivalent domestic asset value. Although the presentation of this paper concentrates on jump diffusion models, the method can be generally applied to Levy processes for the foreign asset and the FX rate. A future work may consider if the results obtained in this paper still hold for path-dependent quanto option pricing under jump diffusion economies. In particular, an interesting extension is about its application in quanto lookback options considered by Dai, Wong and Kwok (2004).

References

1. Dai, M., Wong, H. Y., and Kwok, Y. K., Quanto Lookback Options, *Mathematical Finance*, **14**, (2004), 445–467.
2. Huang, S. C. and Hung, M. W., Pricing Foreign Equity Options under Levy Processes. *Journal of Futures Markets*, **25**, (2005), 917–944.
3. Kou, S. G., A Jump Diffusion Model for Option Pricing. *Management Science*. **48**, (2002), 1086–1101.
4. Kwok, Y. K. and Wong, H. Y., Currency-translated Foreign Equity Options with Path-dependent Features and Their Multi-asset Extensions. *International Journal of Theoretical and Applied Finance*, **3**, (2000), 257–278.
5. Merton, R. C., Option Pricing When Underlying Stock Returns Are Discontinuous. *Journal of Financial Economics*, **3**, (1976), 125–144.
6. Reiner, E., Quanto Mechanics. *Risk*, **5**, (1992), 59–63.

7. Toft, K. B. and Reiner, E. S., Currency-translated Foreign Equity Options: The American Case. *Advances in Futures and Options Research*, **9**, (1997), 233–264.
8. Wong, H. Y. and Lau, K. Y., Analytical Valuation of Turbo Warrants under Double Exponential Jump Diffusion. *Journal of Derivatives* Summer, (2008), 61–73.